EARTH DEMOCRACY

EARTH DEMOCRACY

Justice, Sustainability, and Peace

VANDANA SHIVA

North Atlantic Books
Berkeley, California

Published by
North Atlantic Books
Berkeley, California

Cover photo © Jakkree Thampitakkull/Shutterstock.com
Cover design by Jasmine Hromjak
Book design by Adept Content Solutions
Printed in the United States of America

First published in 2005 by South End Press.

Earth Democracy: Justice, Sustainability, and Peace is sponsored and published by North Atlantic Books, an educational nonprofit based in Berkeley, California, that collaborates with partners to develop cross-cultural perspectives, nurture holistic views of art, science, the humanities, and healing, and seed personal and global transformation by publishing work on the relationship of body, spirit, and nature.

North Atlantic Books' publications are distributed to the US trade and internationally by Penguin Random House Publishers Services. For further information, visit our website at www.northatlanticbooks.com.

Library of Congress Cataloging-in-Publication Data

Shiva, Vandana, author.
Earth democracy : justice, sustainability, and peace / Vandana Shiva.
pages cm
Originally published: Cambridge, Mass. : South End Press, c2005.
Includes bibliographical references and index.
Summary: "Calls for a radical shift in the values that govern democracies, condemning the role that unrestricted capitalism has played in the destruction of environments and livelihoods"— Provided by publisher.
ISBN 978-1-62317-041-7 (trade paper) — ISBN 978-1-62317-042-4 (e-book)
1. Social justice. 2. Environmental justice. 3. Sustainable development.
4. Sustainability—Political aspects. I. Title.
HM671.S457 2015
303.3'72—dc23 2015024094

6 7 8 9 10 KPC 24 23 22 21

This book includes recycled material and material from well-managed forests. North Atlantic Books is committed to the protection of our environment. We print on recycled paper whenever possible and partner with printers who strive to use environmentally responsible practices.

CONTENTS

INTRODUCTION

Principles of Earth Democracy

Earth Democracy is both an ancient worldview and an emergent political movement for peace, justice, and sustainability. Earth Democracy connects the particular to the universal, the diverse to the common, and the local to the global. It incorporates what in India we refer to as vasudhaiva kutumbkam (the earth family)—the community of all beings supported by the earth. Native American and indigenous cultures worldwide have understood and experienced life as a continuum between human and nonhuman species and between present, past, and future generations. An 1848 speech attributed to Chief Seattle of the Suquamish tribe captures this continuum.

> How can you buy or sell the sky, the warmth of the land? The idea is strange to us.
> If we do not own the freshness of the air and the sparkle of the water, how can you buy them?
> Every part of this earth is sacred to my people. Every shining pine needle, every sandy shore, every mist in the dark woods, every clearing and humming insect is holy in the memory and experience of my people. The sap which courses through the trees carries the memories of the red man.
> This we know; the earth does not belong to man; man belongs to the earth.
> This we know. All things are connected like the blood which unites our family. All things are connected.

Earth Democracy is the awareness of these connections and of the rights and responsibilities that flow from them. Chief Seattle's protest that "the earth does not belong to man" finds echoes across the world:

"Our world is not for sale," "Our water is not for sale," "Our seeds and biodiversity are not for sale." This response to privatization under the insane ideology known as corporate globalization builds Earth Democracy. Corporate globalization sees the world only as something to be owned and the market only as driven by profit. From Bangalore in 1993, when half a million Indian peasants pledged to resist the classification of seeds as private property required by the Trade Related Aspects of Intellectual Property Rights (TRIPS) agreement of the World Trade Organization (WTO) to Seattle in 1999 and Cancun in 2003 when protests stopped the WTO ministerial meetings, the corporate globalization agenda has been responded to creatively, imaginatively, and courageously by millions of people who see and experience the earth as a family and community consisting of all beings and humans of all colors, beliefs, classes, and countries.

In contrast to viewing the planet as private property, movements are defending, on a local and global level, the planet as a commons. In contrast to experiencing the world as a global supermarket, where goods and services are produced with high ecological, social, and economic costs and sold for abysmally low prices, cultures and communities everywhere are resisting the destruction of their biological and cultural diversity, their lives, and their livelihoods. As alternatives to the suicidal, globalized free market economy based on plundering and polluting the earth's vital resources, which displaces millions of farmers, craftspeople, and workers, communities are resolutely defending and evolving living economies that protect life on earth and promote creativity.

Corporate globalization is based on new enclosures of the commons; enclosures which imply exclusions and are based on violence. Instead of a culture of abundance, profit-driven globalization creates cultures of exclusion, dispossession, and scarcity. In fact, globalization's transformation of all beings and resources into commodities robs diverse species and people of their rightful share of ecological, cultural, economic, and political space. The "ownership" of the rich is based on the "dispossession" of the poor—it is the common, public resources of the poor which are privatized, and the poor who are disowned economically, politically, and culturally.

Patents on life and the rhetoric of the "ownership society" in which everything-water, biodiversity, cells, genes, animals, plants—is property express a worldview in which life forms have no intrinsic worth, no integrity, and no subjecthood. It is a worldview in which the rights of farmers to seed, of patients to afford able medicine, of producers to a fair share of nature's resources can be freely violated. The rhetoric of the "ownership society" hides the anti-life philosophy of those who, while mouthing pro-life slogans, seek to own, control, and monopolize all of the earth's gifts and all of human creativity. The enclosures of the commons that started in England created millions of disposable people. While these first enclosures stole only land, today all aspects of life are. being enclosed—knowledge, culture, water, biodiversity, and public services such as health and education. Commons are the highest expression of economic democracy.

The privatization of public goods and services and the commoditization of the life support systems of the poor is a double theft which robs people of both economic and cultural security. Millions, deprived of a secure living and identity, are driven toward extremist, terrorist, fundamentalist movements. These movements simultaneously identify the other as enemy and construct exclusivist identities to separate themselves from those with whom, in fact, they are ecologically, culturally, and economically connected. This false separation results in antagonistic and cannibalistic behavior. The rise of extremism and terrorism is a response to the enclosures and economic colonization of globalization. Just as cannibalism among factory-farmed animals stops when chicken and pigs are allowed to roam free, terrorism, extremism, ethnic cleansing, and religious intolerance are unnatural conditions caused by globalization and have no place in Earth Democracy.

Enclosures create exclusions, and these exclusions are the hidden cost of corporate globalization. Our movements against the biopiracy of neem, of basmati, of wheat have aimed at and succeeded in reclaiming our collective biological and intellectual heritage as a commons. Movements such as the victorious struggle started by the tribal women of a tiny hamlet called Plachimada in India's Kerala state against one of the world's largest corporations, Coca-Cola, are at the heart of the emerging Earth Democracy.

New intellectual property rights enclose the biological, intellectual, and digital commons. Privatization encloses the water commons. The enclosure of each common displaces and disenfranchises people which creates scarcity for the many, while generating "growth" for the few. Displacement becomes disposability, and in its most severe form, the induced scarcity becomes a denial of the very right to live. As the use of genetically modified seed and sex-selective abortions spread, large groups of people—especially women and small farmers—are disappearing. The scale and rate of this disappearance is proportional to the "economic growth" driven by the forces of neoliberal corporate globalization. However, these brutal extinctions are not the only trend shaping human history.

On the streets of Seattle and Cancun, in homes and farms across the world, another human future is being born. A future based on inclusion, not exclusion; on nonviolence, not violence; on reclaiming the commons, not their enclosure; on freely sharing the earth's resources, not monopolizing and privatizing them. Instead of being shaped by closed minds behind closed doors, as the hawkish right-wing Project for the New American Century was, the people's project is unfolding in an atmosphere of dialogue and diversity, of pluralism and partnerships, and of sharing and solidarity. I have named this project Earth Democracy. Based on our self-organizing capacities, our earth identities, and our multiplicities and diversity, Earth Democracy's success concerns not just the fate and well-being of all humans, but all beings on the earth. Earth Democracy is not just about the next protest or the next World Social Forum; it is about what we do in between. It addresses the global in our everyday lives, our everyday realities, and creates change globally by making change locally. The changes may appear small, but they are far-reaching in impact—they are about nature's evolution and our human potential; they are about shifting from the vicious cycles of violence in which suicidal cultures, suicidal economies, and the politics of suicide feed on each other to virtuous cycles of creative nonviolence in which living cultures nourish living democracies and living economies.

Earth Democracy is not just a concept, it is shaped by the multiple and diverse practices of people reclaiming their commons, their resources, their livelihoods, their freedoms, their dignity, their identities, and their peace. While these practices, movements, and actions are multifaceted and multiple, I have tried to identify clusters that present the ideas and examples of living democracies, living cultures, and live economies which together build Earth Democracy. Economy, politics, culture are not isolated from each other. The economies through which we produce and exchange goods and services are shaped by cultural values and power arrangements in society. The emergence of living economies, living cultures, and living democracies is, therefore, a synergistic process.

Living economies are processes and spaces where the earth's resources are shared equitably to provide for our food and water needs and to create meaningful livelihoods. Earth Democracy evolves from the consciousness that while we are rooted locally we are also connected to the world as a whole, and, in fact, to the entire universe. We base our globalization on ecological processes and bonds of compassion and solidarity, not the movement of capital and finance or the unnecessary movement of goods and services. A global economy which takes ecological limits into account must necessarily localize production to reduce wasting both natural resources and people. And only economies built on ecological foundations can become living economies that ensure sustainability and prosperity for all. Our economies are not calculated in the short term of corporate quarterly returns or the four- to five-year perspective of politicians. We consider the evolutionary potential of all life on earth and re-embed human welfare in our home, our community, and the earth family. Ecological security is our most basic security; ecological identities are our most fundamental identity. We are the food we eat, the water we drink, the air we breathe. And reclaiming democratic control over our food and water and our ecological survival is the necessary project for our freedom.

Living democracy is the space for reclaiming our fundamental freedoms, defending our basic rights, and exercising our common responsibilities and duties to protect life on earth, defend peace, and promote justice. Corporate globalization promised that free markets would promote democracy. On the contrary, the free markets of global corporations have destroyed democracy at every level. At the most fundamental level, corporate globalization destroys grassroots democracy through the enclosure of the commons. The very rules of globalization, whether imposed by the World Bank and the International Monetary Fund (IMF) or by the WTO, have been written undemocratically, without the participation of the most affected countries and communities. Corporate globalization undermines and subverts national democratic processes by taking economic decisions outside the reach of parliaments and citizens. No matter which government is elected, it is locked into a series of neoliberal reform policies. Corporate globalization is in effect the death of economic democracy. It gives rise to corporate control and economic dictatorship.

When economic dictatorship is grafted onto representative, electoral democracy, a toxic growth of religious fundamentalism and right-wing extremism is the result. Thus, corporate globalization leads not just to the death of democracy, but to the democracy of death, in which exclusion, hate, and fear become the political means to mobilize votes and power.

Earth Democracy enables us to envision and create living democracies. Living democracy enables democratic participation in all matters of life and death—the food we eat or do not have access to; the water we drink or are denied due to privatization or pollution; the air we breathe or are poisoned by. Living democracies are based on the intrinsic worth of all species, all peoples, all cultures; a just and equal sharing of this earth's vital resources; and sharing the decisions about the use of the earth's resources.

Living cultures are spaces in which we shape and live our diverse values, beliefs, practices, and traditions, while fully embracing our common, universal humanity, and our commonality with other species through soil, water, and air. Living cultures are based on nonviolence and compassion, diversity and pluralism, equality and justice, and respect for life in all its diversity.

Living cultures that grow out of living economies have space for diverse species, faiths, genders, and ethnicities. Living cultures grow from the earth, emerging from particular places and spaces while simultaneously connecting all humanity in a planetary consciousness of being members of our earth family. Living cultures are based on multiple and diverse identities. They are based on earth identity as both the concrete reality of our everyday lives—where we work, play; sleep, eat, laugh, or cry—and the processes which connect us globally.

"All things are connected," Chief Seattle tells us. We are connected to the earth locally and globally. Living cultures based on the recovery of our earth identity create the potential for reintegrating human activities into the earth's ecological processes and limits. Remembering we are earth citizens and earth children can help us recover our common humanity and help us transcend the deep divisions of intolerance, hate, and fear that corporate globalization's ruptures, polarization, and enclosures have created.

Ancient concepts of living peacefully as one while evolving in biologically and culturally diverse trajectories cross our interconnected earth family and inspire Earth Democracy. The ancient wisdom and tradition of nonseparability and interconnectedness that we revive is evident in quantum theory; the space-time continuum of general relativity, and the self-organized complexity of living organisms.

Earth Democracy, in the contemporary context, reflects the values, worldviews, and actions of diverse movements working for peace, justice, and sustainability. We live in times when the combination of representative democracy and economic globalization has generated new fears, new insecurities, new fundamentalisms, and new violence. The 2004 elections, both in India and in the US, show how in the face of job loss and the destruction of livelihoods a fundamentalist religious discourse fills the space. This discourse polarizes society and allows cultural differences to be used as a wedge to divide people from the issues that tie them together—their jobs, the environment, human rights, and one common humanity.

Earth Democracy allows us to reclaim our common humanity and our unity with all life. Earth Democracy relocates the sanctity of life in all beings and all people irrespective of class, gender, religion, or caste. And it redefines "upholding family values" as respecting the limits on greed and violence set by belonging to the earth family. Family values of the earth family do not allow for the privatization of water or the patenting of life, since all beings have a right to life and well-being. In the earth family that acknowledges, as Chief Seattle did, that "all things share the same breath, the beast, the tree, the man.... The air shares its spirit with all the life it supports," one part of the international community cannot destabilize the climate, enclose the atmospheric commons, or ignore the rights of other species and other countries by creating 36 percent of the world's CO_2 pollution.

Earth Democracy protects the ecological processes that maintain life and the fundamental human rights that are the basis of the right to life, including the right to water, the right to food, the right to health, the right to education, and the right to jobs and livelihoods. Earth Democracy is based on the recognition of and respect for the life of all species and all people.

Over the past three decades, my conception of Earth Democracy has been concretized through my engagement with diverse movements. Ecology movements, conservation movements, and animal rights movements have based their struggles around the intrinsic worth and value of all species. Human rights movements have been rooted in the recognition of the universal human rights of all people. In Earth Democracy, the concern for human and nonhuman species comes together in a coherent, nonconflicting whole that provides an alternative to the worldview of corporate globalization, which gives rights only to corporations and which sees human and other beings as exploitable raw material or disposable waste.

Earth Democracy connects us through the perennial renewal and regeneration of life—from our daily life to the life of the universe. Earth Democracy is the universal story of our times, in our different places. It pulsates with the limitless potential of an unfolding universe even while it addresses the real threats to our very· survival as a species. It is hope in a time of hopelessness, it brings forth peace in a time of wars without end, and it encourages us to love life fiercely and passionately at a time when leaders and the media breed hatred and fear.

Principles of Earth Democracy

1. All species, peoples, and cultures have intrinsic worth

All beings are subjects who have integrity, intelligence, and identity, not objects of ownership, manipulation, exploitation, or disposability. No humans have the right to own other species, other people, or the knowledge of other cultures through patents and other intellectual property rights.

2. The earth community is a democracy of all life

We are all members of the earth family, interconnected through the planet's fragile web of life. We all have a duty to live in a manner that protects the earth's ecological processes, and the rights and welfare of all species and all people. No humans have the right to encroach on the ecological space of other species and other people, or to treat them with cruelty and violence.

3. Diversity in nature and culture must be defended

Biological and cultural diversity is an end in itself. Biological diversity is a value and a source of richness, both materially and culturally that creates conditions for sustainability. Cultural diversity creates the conditions for peace. Defending biological and cultural diversity is a duty of all people.

4. All beings have a natural right to sustenance

All members of the earth community, including all humans, have the right to sustenance—to food and water, to a safe and dean habitat, to security of ecological space. Resources vital to sustenance must stay in the commons. The right to sustenance is a natural right because it is the right to life. These rights are not given by states or corporations, nor can they be extinguished by state or corporate action. No state or corporation has the right to erode or undermine these natural rights or enclose the commons that sustain life.

5. Earth Democracy is based on living economies and economic democracy

Earth Democracy is based on economic democracy. Economic systems in Earth Democracy protect ecosystems and their integrity; they protect people's livelihoods and provide basic needs to all. In the earth economy there are no disposable people or dispensable species or cultures. The earth economy is a living economy. It is based on sustainable, diverse, pluralistic systems that protect nature and people, are chosen by people, and work for the common good.

6. Living economies are built on local economies

Conservation of the earth's resources and creation of sustainable and satisfying livelihoods are most caringly, creatively, efficiently, and equitably achieved at the local level. Localization of economies is a social and ecological imperative. Only goods and services that cannot be produced locally—using local resources and local knowledge—should be produced nonlocally and traded long distance. Earth Democracy is based on vibrant local economies, which support national and global economies. In Earth Democracy, the global economy does not destroy and crush local economies, nor does it create disposable people. Living economies recognize the creativity of all humans and create spaces for diverse creativities to reach their full potential. Living economies are diverse and decentralized economies.

7. Earth Democracy is a living democracy

Living democracy is based on the democracy of all life and the democracy of everyday life. In living democracies people can influence the decisions over the food we eat, the water we drink, and the health care and education we have. Living democracy grows like a tree, from the bottom up. Earth Democracy is based on local democracy, with local communities—organized on principles of inclusion, diversity, and ecological and social responsibility—having the highest authority on decisions related to the environment and natural resources and to the sustenance and livelihoods of people. Authority is delegated to more distant levels of governments on the principle of subsidiarity. Self-rule and self-governance is the foundation of Earth Democracy.

8. Earth Democracy is based on living cultures

Living cultures promote peace and create free spaces for the practice of different religions and the adoption of different faiths and identities. Living cultures allow cultural diversity to thrive from the ground of our common humanity and our common rights as members of an earth community:

9. Living cultures are life nourishing

Living cultures are based on the dignity of and respect for all life, human and nonhuman, people of all genders and cultures, present and future generations. Living cultures are, therefore, ecological cultures which do not promote life-destroying lifestyles or consumption and production patterns, or the overuse and exploitation of resources. Living cultures are diverse and based on reverence for life. Living cultures recognize the multiplicity of identities based in an identity of place and local community—and a planetary consciousness that connects the individual to the earth and all life.

10. Earth Democracy globalizes peace, care, and compassion

Earth Democracy connects people in circles of care, cooperation, and compassion instead of dividing them through competition and conflict, fear and hatred. In the face of a world of greed, inequality, and overconsumption, Earth Democracy globalizes compassion, justice, and sustainability.

CHAPTER ONE

Living Economies

> The earth provides enough resources for everyone's need, but not for some people's greed.
>
> —Mahatma Gandhi[1]

The word *economics* is derived from the Greek word *oikos*, meaning home. Home is where one is born, grows up, and is looked after. Mathew Fox wrote, "Our true home is the universe itself."[2] Robert Frost observes, "Home is the place where when you go there, they have to take you in."[3] Home is where there is always a place for you at the table and where you can count on sharing what is at the table. To be part of a home, a household, is to have access to life. How is it that economic systems today are such unwelcoming spaces? How have they become places that, rather than take us in, often bar our entry, and, in the process, refuse us not only a home, but a right to sustenance, stability, and ultimately life?

The dominant economy goes by many names—the market economy, the globalized economy, corporate globalization, and capitalism, to name a few—but all these names fail to acknowledge that this economy is but one of the three major economies at work in the world today. In Earth Democracy every being has equal access to the earth's resources that make life possible; this access is assured by recognizing the importance of the other two economies: nature's economy and the sustenance economy.

The globalized free market economy, which dominates our lives, is based on rules that extinguish and deny access to life and livelihoods by generating scarcity. This scarcity is created by the destruction of nature's economy and the sustenance economy, where life is nourished, maintained, and renewed. Globalization and free trade decimate the conditions for productive, creative employment by enclosing the commons, which are necessary for the sustenance of life. The anti-life dimensions of economic globalization are rooted in the fact that capital exchange is taking the place of living processes and the rights of corporations are displacing those of living people.

The economic conflict of our times is not just a North-South divide, though the inequalities created by colonialism, the mal development model of debt-slavery imposed by the IMF and the World Bank, and the rules of the WTO—have that dimension. The contest is between a global economy of death and destruction and diverse economies for life and creation. In our age, "have or have not" has mutated into "live or live not."

The Three Economies

Why does the dominant economic model fail to meet the needs of so many societies and communities? Why is success, measured by economic growth, so intimately related to increased poverty, hunger, and thirst? There are two reasons why ecological disasters and the number of displaced, destitute, and disposable people increase in direct proportion to economic growth. First is the reduction of the visible economy to the market and activities controlled by capital. Second, the legal rights of corporations have increased at the cost of the rights of real people.

As the dominant economy myopically focuses on the working of the market, it ignores both nature's economy and the sustenance economy, on which it depends. In a focus on the financial bottom line, the market makes invisible nature's economy and people's sustenance economies.

While the exchange of goods and services has always been a characteristic of human societies, the elevation of the market to the highest organizing principle of society has led to the neglect of the other two vital economies. When exclusive attention is given to the growth of the market, living processes become invisible externalities. The requirements of nature, not backed up by suit able purchasing power, cannot be registered or fulfilled by the market economy. Not only does this focus on the market hide the existence of nature's economy and the sustenance economy, it hides the harm that market growth causes. As a result, especially in the context of Third World countries, nature's need for resources gets ignored. So too are the requirements of the sustenance economy that provides for the biological survival of the marginalized poor and the reproduction of society. The hidden damage. caused by market-based development and globalization processes have created new forms of poverty and underdevelopment. The political economy of Earth Democracy movements cannot be understood without a dear comprehension of the place of natural resources in the three distinct economies.

Furthermore, economic growth is leading to disenfranchisement and disposability of people as hard-won rights designed to protect people are increasingly transferred to corporations. Free market rules for corporate freedom are increasingly rules which exclude real people from the economic and political affairs of society and disenfranchise them from nature. The legal status of corporations serves to hold the people running the corporations free from responsibility for the harm the corporations cause. And just as the corporation gained the legal standing of a person, the market, too, has become personified. More pages in the media are devoted to the "health of the market" than to the health of the planet or well-being of people.

Nature's Economy

Like the word *economy, ecology* also comes from the Greek word for home or household. Yet in the context of market-oriented development, economy and ecology have been pitted against each other. The market economy separates nature from people and ecology from economy. Nature is defined as free of humans. Conservation is reduced to "wilderness" management. Development is viewed as the exclusive domain of production. Nature and people's self-provisioning economies have no productive role according to the market.

However nature's economy is the first economy, the primary economy on which all other economies rest. Nature's economy consists of the production of goods and services by nature-the water recycled and distributed through the hydrologic cycle, the soil fertility produced by microorganisms, the plants fertilized by pollinators. Human production, human creativity shrinks to insignificance in comparison with nature.

Natural resources are produced and reproduced through a complex network of ecological processes. Nature is the world's dominant producer, but its products are not, and cannot be, acknowledged as such in the market economy. Only production and productivity in the context of market economics has been considered production. Organic productivity in forestry or agriculture has also been viewed narrowly with the marketable products as the total productive output. This has resulted in vast areas of productive work—the production of humus by forests, the regeneration of water resources, the natural evolution of genetic products, the creation of fertile soil from eroding rock—remaining beyond the scope of economics. Many of these productive processes are dependent on a number of other ecological processes and are not fully understood even within the natural sciences.

At present, ecology movements are the primary voice stressing the full economic value of these natural processes. The present scale of resource-ignorant economic development threatens the whole natural resource system with a serious loss of productivity. The marker-oriented development process risks destroying nature's economy through the overexploitation of resources and the uncomprehended destruction of ecological processes. These consequences do not necessarily manifest themselves within the period of a given development project. The short-term positive contribution of economic growth from such development may prove totally inadequate to balance the invisible or delayed damage to the economy of natural ecological processes. In the larger context, economic growth can thus, itself become the source of underdevelopment. The ecological destruction associated with uncontrolled exploitation of natural resources for commercial gains is symptomatic of the conflict over how wealth is generated in the market and in nature.

The Sustenance Economy: Bringing People Back into the Picture

In the sustenance economy, people work to directly provide the conditions necessary to maintain their lives. This is the economy through which human production and reproduction is primarily possible. It is the women's economy where, because of the patriarchal division of labor, societal reproduction takes place. Women's work provides sustenance and support to all human activities—including the visible activities of the market dominated economy. The sustenance economy is the economy of the two-thirds of humanity engaged in craft production, peasant agriculture, artisanal fishing, and indigenous forest economies. The sustenance economy includes all spheres in which humans produce in balance with nature and reproduce society through partnerships, mutuality, and reciprocity.

Without the sustenance economy, there would be no market economy. Sustenance economies exist even where capital markets do not. Yet capital's markets cannot exist without the sustenance economy, nor can the market fully internalize the sustenance economy because externalizing the social burden is the very basis of profits and capital accumulation. As structural adjustments and globalization destroy livelihoods, women work longer hours, at multiple part-time jobs to feed their families. As privatization of health care dismantles public health systems, families take on the burden of looking after the ill. The more markets depend on work outside the market, the more the sustenance economy is rendered invisible, and left without resources.

The poverty of the Third World has resulted from centuries of the drain of resources from the sustenance economy. Globalization has accelerated and expanded the methods used to deplete the sustenance economy—the privatization of water, the patenting of seeds and biodiversity, and the corporatization of agriculture. This deliberate starving of the sustenance economy is at the root of the violence of globalization.

Modern economics, the concept of development and progress, and now the paradigm of globalization cover but a minuscule portion of the history of human economic production. The sustenance economy has given human societies the material means of survival by deriving livelihoods directly from nature. Within the context of a limited resource base, diverting natural resources from directly sustaining human existence to generating growth in the market economy destroys the sustenance economy. In the sustenance economy, satisfying basic needs and ensuring long-term sustainability are the organizing principles for natural resource use, whereas the exploitation of resources for profits and capital accumulation are the organizing principles for the market.

Markets and the Market

Markets are places of exchange. The bazaar, thriving even today in India, is a place where people exchange products they have grown and produced. The concrete, embedded market grows out of society. Based on direct relationships and face-to-face transactions, it is in fact an extension of society. When markets are replaced by *the market,* society is replaced by capital and the market becomes the anonymous face of corporations. Real people, exchanging what they create and what they need, are replaced by the abstract and invisible hand of the market.

There are two kinds of markets. Markets embedded in nature and society are places of exchange, of meeting, of culture. Some are simultaneously cultural festivals and spaces for economic transactions, with real people buying and selling real things they have produced or directly need. Such markets are diverse and direct. They serve people, and are shaped by people.

The market shaped by capital, excludes people as producers. Cultural spaces of exchange are replaced by invisible processes. People's needs are substituted for by greed, profit, and consumerism. The market becomes the mystification of processes of crude capital accumulation, the mask behind which those wielding corporate power hide.

It is this disembodied, decontextualized market which destroys the environment and peoples' lives.

The Domination of the Market

A key to the domination of the market economy is its ability to claim resources from outside of its scope. The transformation of land from public to private ownership was essential for the market economy to become the dominant economy. The transformation, known as the enclosure of the commons, was usually triggered by the greed and power of the privateers.[4] The word *enclosure* describes the physical exclusion of the community from their commons by "surrounding a piece of land with hedges, ditches, or other barriers to the free passage of men and animals."[5]

The land called the commons was formally owned by the landlord, but the rights to use it belonged to the commoners. It was the removal of these rights to common property which enabled enclosure. In England, where the movement began in the sixteenth century, the enclosures were driven by the hunger of machines, by the increased demand for wool by the textile industry. The landlords, supported by industrialists, merchants, and bankers, pushed peasants off the land and replaced them with sheep. "Sheep eat men" is how Sir Thomas More described the phenomena of the enclosures of the commons.

> Your sheep, that were wont to be so meek and tame and so small eaters, now, and I had hearsay become so great devourers and so wild, that they eat up and swallow down the very man themselves. They consume, destroy and devour whole fields, houses and cities.[6]

The economics of enclosure worked for the landlords, but against the peasants. While one acre of arable land on the commons could produce 670 pounds of bread, it could only maintain a few sheep.[7] In terms of the food and the sustenance economy this was a loss, since the sheep could only produce 176 pounds of mutton. However, in monetary terms, the landlords gained. A single shepherd tending the sheep produced much higher returns for the landlord than the rents paid by dozens of peasants. That the peasants were growing the food, fodder, fuel, and other essentials for their survival on the commons was, for the landlord, not a matter of consideration. The profits provided the necessary justification for the privateers to expand the market economy, despite the cost to nature's economy and the sustenance economy.

Five processes constitute enclosing the commons.

1. The exclusion of people from access to resources that had been their common property or held in common.

2. The creation of "surplus" or "disposable" people by denying rights of access to the commons that sustained them.

3. The creation of private property by the enclosure of common property.

4. The replacement of diversity that provides for multiple needs and performs multiple functions with monocultures that provide raw material and commodities for the market.

5. The enclosure of minds and imagination, with the result that enclosures are defined and perceived as universal human progress, not as growth of privilege and exclusive rights for a few and dispossession and impoverishment for the many.

Enclosures were exalted as allowing "an unparallel expansion of productive possibilities."[8] Productivity was defined from the perspective of the rich and the powerful, not from that of the commoner, and valued only profits and the benefit to the market, not nature's sustainability or people's sustenance. The rich "deplored the insubordination of commoners, the unimprovability of their pastures, and the brake on production represented by shared property."[9] Despite the opinion of the landlords, the commons were not wasted land; they were a rich resource providing the community with a degree of self-reliance and self-governance.

The market created its own multiplier effect, which pushed the sustenance economy further from view. The more the powerful gained economic and political power from the growing market economy, the more they dispossessed the poor and enclosed their common property. And the more the poor were dispossessed of their means to provide their own sustenance, the more they had to turn to the market to buy what they had formerly produced themselves.

> When the cottager was cut off from his resources . . . there was little else that he could do in the old way. It was out of the question to obtain most of his supplies by his own handiwork: they had to be procured, ready made from another source. That source, I need hardly say, was a shop.[10]

As it was with the land commons, so it is today with the biodiversity and seed commons through intellectual property rights and the water commons through privatization. Now the seeds, the medicine, the water that historically have been the common property of communities need to be bought at high cost from gene giants like Monsanto, who own the patents, and water giants like Suez, Bechtel, and Vivendi, who own the concessions. The transformation of common property rights into private property rights implicitly denies the right to survival for large sections of society.

What are the Commons?

The very notion of the commons implies a resource is owned, managed, and used by the community. A commons embodies social relations based on interdependence and cooperation. There are clear rules and principles; there are systems of decision-making. Decisions about what crops to sow, how many cattle will graze, which trees will be cut, which streams will irrigate which field at what time, are made jointly and democratically by the members of the community. A democratic form of governance is what made, and makes, a commons a commons. This was as true of England in the late eighteenth century as it is of regions where community control of the commons is still the method of governance and ownership.

In India, the equitable distribution of land was frequently based on a system called *bhaiachara* or custom *(achara)* of the brothers *(bhai)*. Lord Baden-Powell observed in his book on land tenures:

> Here the whole area available was studied and was classified by the Panchayat [local or community council] into good and bad, better, best, &c.; and then a suitable number of lots were made,each consisting of specimen strips of each kind of soil, scattered over the whole area. Each lot so made up would be called the baiwadi-bigha, or tauzi-bigha—an artificial land unit, which had no relation to the ordinary or standard measure; then, ac cording to the requirement of the numbers in the families, a certain number of such units would be handed over to each section and subsection.... Whatever was done, it was always with the desire of equality—adjusting the share to the burden to be borne.[11]

While not all bhaiachara villages went in for strip farming, the practice was most evident in regions where the weather and environment were most severe and unpredictable. Scattering promoted cooperation and collective action.

> Scattering was an institutional device to provide insurance to individuals against uncertainty, provided they cooperated.... The propensity to act collectively increased if risk could be shared and if access to varied resources could be equalized. Scattering of the arable land went hand in hand with compact grazing and collective management of the field channels of irrigation wells and ponds.[12]

Chakravarty-Kaul, a professor at Delhi University, writes that the term bhaiachara also applied to villages where land was divided on the basis of how much was ploughed, ancestral usage, or even where it was customary that each family was allowed to cultivate as much of the commons as it could without putting pressures on other members of the community. The common factor of bhaiachara holdings was that taxes were paid only on what was actually cultivated by the family.[13]

Terra Nullius

Most sustainable cultures, in all their diversity, view the earth as *terra mater* (mother earth). They gratefully receive nature's gifts and return the debt through ecologically sustainable lifestyles and earth-centered cosmologies. The colonial construct of the passivity of the earth and the consequent creation of the colonial category of land as *terra nullius* (empty land), served two purposes: it denied the existence and prior rights of original inhabitants and it obscured the regenerative capacity and processes of the earth. It therefore allowed the emergence of private property from enclosures, and allowed nonsustainable use of resources to be considered "development" and "progress." For the privateer and the colonizer, enclosure was improvement.

In Australia, the colonizers justified the total appropriation of land and its natural resources by declaring the entire continent of Australia to be terra nullius—uninhabited. This declaration established a simple path to privatizing the commons, because as far as the colonizers were concerned, there were no commons. Similarly, in the American colonies the takeover of native resources was justified on the ground that indigenous people did not "improve" their land. As John Winthrop, first governor of the Massachusetts Bay Colony, wrote in 1669:

> Natives in New England, they enclose no land, neither have they any settled habitation, nor any tame cattle to improve the land by soe have nor other but a Natural Right to those countries. Soe as if we leave them sufficient for their use, we may lawfully take the rest.[14]

As I wrote in *Biopiracy*, the logic of empty lands is now being expanded to "empty life." Terra nullius is now used to appropriate biodiversity from the original owners and innovators by defining their seeds, medicinal plants, and medical knowledge as nature, and treating the tools of genetic engineering as the only path to "improvement."[15] By disregarding non-market use, authorities free themselves to enclose rivers through India's River Linking Project and to enclose. groundwater for bottled water and soft drinks by corporations like Coca-Cola and Pepsi.

The English Enclosures

The commons, which the Crown of England declared wastelands, were really productive lands providing extensive common pastures for the animals of the established peasant communities; timber and stone for building; reeds for thatching and baskets; wood for fuel; and wild animals, birds, fish, berries, and nuts for food. These areas supported large numbers of small peasants through common rights of access to these resources. They also gave shelter to the poorer and landless peasants who migrated from the overcrowded open-field villages of the corn-growing districts.[16]

The fate of the forests was similar to that of the pastures. The Crown possessed the forests, while the peasants had common rights to what the forest produced. With the increasing demand for resources to feed capitalist growth, the Crown adopted a policy of deforestation. As a result, the peasants lost their common rights and the Crown and the lords of manors enclosed the deforested land and parceled it into large farms for lease at competitive rents.

A head-on clash developed between the lords of manors and the peasantry in many parts of the country over control of the commons. Between 1628 to 1631, large crowds repeatedly attacked and broke down the enclosures. Large areas of England were in a state of rebellion.

Under English common law, enclosure of a common required the unanimous consent of the entire community. No authority had the right to alienate and enclose the commons. Just one member could block the change. This right was fundamental and inalienable, and was fiercely defended.

> I defy you to enclosure one square yard; I defy you severally; I defy you jointly; you may meet in your court, you may pass what resolutions you please, I shall condemn them; for I have a right to put my beast on this land and every part of it; the law gives me this right and the King protects it.[17]

Local democracy and these inalienable rights were, however, eroded as the power of money subverted the management and ownership of the commons. Commercial interests pressured Parliament to pass acts to enclose the commons.

Between 1770 and 1830, 3,280 bills were passed by Parliament to enact the enclosure of the commons. As a result of this legislation:

> 6 million acres of commonly held lands, open fields, meadows, wetlands, forests and unoccupied "waste" lands, until then the domain of the public at large, were put into private hands and subsequently hedged and fenced and farmed and herded and hunted for private gain.[18]

Enclosures without parliamentary approval were nearly as vast. By the period's end more than half the land in England was in private hands.

From Commons to Commodities: Colonialism as Enclosure

Enclosures are not just a historical episode that occurred in England. Enclosures have also been central to the continuing processes of colonization. Colonialism created private property by enclosing the commons and displacing and uprooting the original peoples in the Americas, Africa, and Asia.

The English policy of deforestation and enclosure was replicated in the colonies in India. The first Indian Forest Act, passed in 1865 by the Supreme Legislative Council, authorized the government to declare forests and *benap* (unmeasured lands) as reserved forests. The act marks the beginning of what is called the "scientific management" of forests and resulted in the erosion of both the fertility of the forests and the rights of local people to forest produce. Though technically the forests became state property, forest reservation was in fact an enclosure because it converted a common resource into a commercial one. The state merely mediated the privatization.

When the British established their rule in India, it was estimated that from one-third to one-half of the total area of Bengal Province alone was wasteland. The colonial concept of wastelands was not an assessment of the biological productivity of land but of its revenue-generating capacity. Wasteland was land which did not yield any revenue because it was not farmland, but forest. These lands were taken over by the British government and leased to cultivators to turn them into revenue-generating lands. It was only at the end of the 19th century when forests also became a source of revenue that state forests were no longer called wastelands. Village forests and grazing lands, however, continued to be categorized as wastelands, even though they were vital fuel and fodder resources for the agricultural economy.

The colonial category of wastelands was thus an economic category, but colonial policy also created the category of ecologically wasted lands—land which had lost its biological productivity because of social and government action and inaction. These wasted lands are found in areas demarcated as reserved forests, owned privately by individuals for agricultural use, and common lands shared by communities for fuel and fodder supplies.

In the colonial period, peasants were forced to grow indigo instead of food, salt was taxed to provide revenues for the British military, and forests were enclosed to transform them into state monopolies for commercial exploitation. In the rural areas, these actions gradually eroded the peasants' usufruct or *nistar* rights—their rights to food, fuel, and grazing livestock on the community's common lands. The erosion of peasant communities' rights to their forests, sacred groves, and "wastelands" has been the prime cause of their impoverishment.

Once India's land was usurped, the collection of public revenue became a prime concern of the colonizing powers. Someone needed to be taxed. Before the British came to India produce was taxed but the land itself was not. To collect the tax, the British needed proprietors of land who would collect rents from the cultivators and pass it on to them. How could this be done?

The answer was extremely simple—create landlords. The task of finding the landlords was not too difficult—who better than those who already were used to collect money from the peasants for the state? The *zamindars* formed the majority of the new landlords.

A motley collection of rural overlords in late 18th-century Bengal conveniently and misleadingly went under the single name of zamindar. To compound the confusion, these varied elements in the Bengal countryside bore no resemblance to the village zamindars. The Bengal zamindars encompassed at least four separately identifiable categories: the old territorial heads of principalities, such as the rajas of Tippera and Coach Behar; the great landholding families who paid a fixed land tax and behaved like feudatory chiefs, such as the Rajas of Burdwan, Dinajpur, Rajshahi, Jessore, and Nadia; the numerous families who had held offices for collecting land revenue over a number of generations; and revenue farmers established by the grant of Diwani in 1765. In a bad "case of mistaken identity," Lord Cornwallis, the governor general, by a grand proclamation on March 22, 1793, followed up by a barrage of regulations, conferred the prized private property right in land to this diverse group of rural overlords unified only in name.[19]

The colonial extraction of resources dramatically transformed India's ability to develop local infrastructure. Dharampal, India's preeminent historian, has shown that in pre-British India, 80 to 95 percent of resources were utilized at the local and intermediate levels for maintaining the socio-cultural-economic infrastructure. Only 5 to 20 percent went to a central authority; the rest remained in the local economy to support performing arts, *vaidyas* (indigenous medical practitioners), school teachers, priests, accountants, iron smiths, carpenters, potters, washer men, water managers, and the maintenance of irrigation works. Colonialism reversed this disbursement ratio, with Britain leaving only 10 percent for local infrastructure to sustain the people and taking 90 percent to run the empire.[20]

Birth of Corporations

The emergence of corporations like the East India Company created new instruments of wealth extraction for the investors and new degrees of impoverishment and dispossession for producers. Instead of producers and, production leading trade, trade controlled production. In the end, corporations took over control of production from the labor guilds in England and from future colonies such as India.

The East India Company, one of the earliest corporations, was created by men who controlled capital to finance voyages for colonization. It was ruled by a governor and 24 assistants and had a monopoly on trade with all islands and ports in Asia, Africa, and America from the Cape of Good Hope to the Straits of Magellan.

In the year 1600, the year the East India Company was born, India was not only nourishing Asia with its rice, wheat, sugar, and raw cotton, but was also the industrial workshop of the world, producing a prodigious quantity of cotton to sell in markets spreading from the farthest reaches of the East Indies and South Asia in the east to Europe in the west, and from the shores of the Caspian Sea to the coast of Mozambique and Madagascar.[21] These international trade routes were conquered by the East India Company and the beginnings of an empire were established. In 1717 the East India Company obtained a *firman*, or grant, from Emperor Farrukhsiyar in Delhi, which allowed, among other concessions, in return for an annual tribute of 3,000 rupees, the company to trade customfree throughout the imperial territories. As historian Radha Kamal Mukherjee writes:

> A whole century of activities of Dutch and British pirates, businesses, soldiers, factors and merchants not merely left them in complete monopoly over the trade between the disparate parts of Asia and between Asia and Europe, but also laid the foundation of empire.[22]

This edict was to become the foundation of the British commercial and political policy in India. In January 1757 Bengal fell. The East India Company's traders were no longer "mere merchants," they were now the rulers of India. For this, the victorious Robert Clive of the East India Company received a reward of £234,000.[23]

The East India Company began by importing finished Indian textiles. Later it banned the import of Indian textiles and limited imports to unfinished products. In 1750, Chinese and Indian regions produced 73 percent of the world's textiles. India was the textile factory of the world. The British destroyed the Indian textile industry and then created their own. The industrial myth credits technology as the reason for the growth of British textiles. However, it was tariffs and prohibitions (in WTO language "quantitative restrictions") which led to the growth of the industry in England. The technological innovations followed. As H. H. Wilson, a professor of history at Oxford, wrote:

> It was stated in evidence in 1813 that the cotton and silk goods of India up to this period could be sold in the British market at a price 50 to 60 per cent lower than those fabricated in England. It consequently became necessary to protect the latter by duties of 70 to 80 per cent on their value or by positive prohibition. Had this not been the case, had not such prohibitory duties and decrees existed, the mills of Parsley and Manchester would have stopped in their outset and could hardly have been again set in motion, even by the powers of steam. They were created by the sacrifice of Indian manufacturers.[24]

Before 1771 England did not produce any cotton cloth at all; it neither grew cotton nor possessed spinners that could make cotton yarn of sufficient strength for the warp. The British textile industry began developing by importing plain white calicos. Then, using the methods, processes, and prints from India, the calico industry advanced in England. By 1845 the tables were turned and England dominated the textile trade. Forgetting history, Sidney Smith could write, "The great object for which the Anglo-Saxon race appears to have been created is the making of Calico."[25]

As India disappeared from production, her contributions disappeared from history. The losses caused by "free" trade-textile exports from India were subject to 80 percent tariffs while imports to India had 2.5 percent tariffs-led to the destruction of both India's domestic and export market. By 1846, India the exporter was importing more than 200 million yards of doth from England, as compared with 51 million yards in 1835 and only 800,000 yards in 1814.[26]

Economic Globalization/Corporate Globalization

In the early stages of industrialization, the enclosures movement in England declared the peasantry dispensable and pushed them off the land. Industrialization was brought as "development" to countries of the South. Corporate rule through globalization continue to build upon the foundation that colonialism created and continues to leave behind it a trail of devastation and destruction.

The market economy inevitably produces a major shift in the way rights to resources are perceived. The transformation of commons into commodities has two implications. It deprives the politically weaker groups of their right to survival, which they had through access to commons, and it robs from nature its right to self-renewal and sustainability, by eliminating the social constraints on resource use that are the basis of common property management.

In Third World countries the transformation of natural re sources into commodities has been largely mediated by the state. Though couched in the language of advancing the collective public interest, the state is often a powerful instrument for the privatization of resources. The transformation of forests from village commons to state-reserved forests serves the interests of the private pulp and paper industry by ensuring cheap and regular supply of raw material. Similarly, dams are built with public funds, but they aim to satisfy the energy and water needs of private industry or the irrigation needs of cash-crop cultivation. Credit from public-sector banks is used to finance the private wells or private trawlers of economically powerful groups. Conflicts over natural resources are conflicts over rights.

Corporate globalization has been imposed on us. It represents itself as the ocean everyone needs to be swimming in, an inevitable process to be part of. Corporate globalization, commonly associated with international trade, is something that many people think they can ignore. But corporate globalization is not about goods crossing national boundaries. We've had international trade forever. Goods were traded across borders before colonialism. In fact, colonialism was caused by the desire to control that trade. Spices were traded long before European colonizers sought control over the spice trade. Corporate globalization now crosses borders with far more serious consequences for the planet and for humanity than those artificial borders that define nations. Compared to the ethical borders being crossed, international trade is very easy to deal with. These ethical boundaries have been created over centuries by faiths, cultures, and societies which declared certain things are not part of commerce. Certain things are not tradable and will be governed by values other than commodity values.

Globalization is, in fact, the ultimate enclosure—of our minds, our hearts, our imaginations, and our resources. Until corporate globalization claimed the resources of this planet—especially water and biodiversity—to be tradable commodities, it was recognized that water couldn't belong to anyone. Rain falls, flows through river basins and underground aquifers, meets the ocean, and evaporates in an amazing hydrological cycle that brings us water. Sometimes the cycle is slow and gives us drought, but we can deal with the drought that the water cycle gives. We cannot deal with engineered drought that says water will only flow one way-uphill to money.

We were promised that globalization would bring us peace by constructing a global village in which everyone would be connected. But the number of wars that have occurred since 1995, when corporate globalization became, literally, the legal constitution of the world, gives lie to this claim. Look at the misunderstandings between cultures that have resulted. In Chapter Three, I talk more about the connections between corporate globalization and the rise of terrorism, the rise of extremism, and the rise of the right wing. A second promise was prosperity: "When the waters rise, all boats will rise." The waters haven't risen. They have fallen. They have been depleted by the very processes of giving control over these resources to corporations.

While unbridled capitalist greed has been referred to as "compassionate capitalism" in the US, compassionate economics of sustenance and nature are precisely what is destroyed by corporate rule and the rule of capital. Protection of nature and people's rights are defined as protectionism, as trade barriers, and as barriers to investment. Trade rules and neoliberal reform institutionalize laws which render compassion itself illegal. Thus, cultures of compassion which treat all life as sacred are made illegal through patents on life. Cultures of compassion and social justice which share social wealth and nature's wealth are rendered illegal through privatization of essential public services like water, health, education. Economies that aim to guarantee and protect livelihoods, jobs, and social security are dismantled, leaving people with no place in society or the economy. These are not examples of compassionate economies, they are examples of a violent economy which looks more and more like warfare, both in its methods and its results.

A typical example of the neoliberal paradigm that dominates current economic and social policy can be found in a book published by the IMF entitled *Who Will Pay? Coping with Aging Societies, Climate Change, and Other Long-Term Fiscal Challenges.* In one fell swoop the book's tide redefines the challenges of social and. ecological reproduction and renewal as fiscal challenges. Humans and nature have disappeared only to be substituted for money. Market economists who see only markets and money are blind to nature and society. They are unable to see that the wealthy's riches are accumulated by exploitation of nature and society. These economists fail to value what nature and people have already contributed. Their contribution is an ecological and social loan greater than any given by the IMF.

Native American and Indian cultures both uphold a seventh-generation logic that states that the impact on the seventh generation should be taken into account before acting. The neoliberal recipe to dismantle Social Security in the present for future generation's security is not really based on the consideration of the welfare of future generations. It is a means of diverting finances from the sustenance economy to the market economy, leaving both present and future generations without security.

Growth and Efficiency in the Market

Another myth about the market economy is that it is the most efficient method of production. But the "efficient" market economy becomes highly inefficient when the destruction of nature's economy is taken into account. The efficiency and productivity of industrial agriculture hides the costs of depletion of soils, exploitation of groundwater, erosion, and extinction of biodiversity. Industrial agriculture uses 10 times more energy than it produces. It uses 10 times more water than biodiverse farming with water-prudent crops and organic practices use. In fact, when assessed from nature's economy, biodiverse, ecological farms have much higher productivity than large-scale, industrial, monoculture farms. The illusion of efficiency is produced by externalizing the ecological costs.[27]

Frequently, growth is generated by converting resources from nature's economy into market commodities. Economic growth takes place through the exploitation of natural resources. Deforestation creates growth. Mining of groundwater creates growth. Overfishing creates growth. Further economic growth cannot help regenerate the very spheres which must be destroyed for economic growth to occur. Nature shrinks as capital grows. The growth of the market cannot solve the very crisis it creates. Furthermore, while natural resources can be converted into cash, cash cannot be converted into nature's wealth. Market economists trying to address the ecological crisis limit themselves to the market, and look for substitutes for the commercial function of natural resources as commodities and raw material. The increased availability of financial resources cannot regenerate the life lost in nature through ecological destruction. An African peasant captured this essence: "You cannot turn a calf into a cow by plastering it with mud."[28] In nature's economy and the sustenance economy the currency is not money, it is life.

Globalized Agriculture

Cash-crop production and food processing divert land and water resources away from sustenance needs and exclude increasing numbers of people from their entitlement to food:

> The inexorable processes of agriculture—industrialization and internationalization—are probably responsible for more hungry people than both cruel wars and unusual whims of nature. There are several reasons why the high-technology-export-crop model increases hunger. Scarce land, credit, water and technology are pre-empted for the export market. Most hungry people are not affected by the market at all.... The profits flow to corporations that have no interest in feeding hungry people without money.[29]

At no point has the global trade of agricultural commodities been assessed in the light of the new conditions of scarcity and poverty that it has induced. This new poverty is no longer cultural and relative, it is absolute and threatening the very survival of millions on this planet. At the root of this new material poverty lies the economic paradigm of the market, which can neither assess the extent of its own requirements for natural resources, nor the impact of this demand on ecological stability and survival. As a result, efficient and productive economic activities within the limited context of the market economy reveal themselves to be inefficient and destructive in the context of the other two economies of nature and sustenance.

In India, the promise of genetically engineered cotton was that it would yield 1,500 kilograms per acre. In four states, the average yield was 200 kilograms. Farmer incomes were projected to increase by 10,000 rupees an acre, but ran to losses of 6,000 rupees an acre. The performance of these crops has been completely unreliable. The hybrid maize seeds that Monsanto sold to peasants in the poorest states of India, like Bihar, left them with total crop failure and losses totaling 4 billion rupees. In the case of the failure of Bt cotton in Andhra Pradesh, it was a billion rupees. A peasant switching to hybrid or genetically modified (GM) seed finds him or herself, in a year's time, two to three hundred thousand rupees in debt. When one company controls the trade, controls the chemicals, controls the market, it sells costly seeds and turns peasants into their biggest buyers. It can only do that by misleading advertising and false projections. What farmers and peasants are left with are very high levels of debt.[30]

It's seed freedom for the corporations but seed slavery for the peasants. Monsanto is selling seeds in India at the same price as in the US. Costs of production have increased tenfold, while prices of agricultural products have plummeted 50 percent because of trade liberalization. Just in food crops alone Indian farmers are losing $24 billion every year. Every year. The poor who were supposed to be made rich are actually finding themselves deeper in poverty. The collapse of rural incomes erodes purchasing power and entitlements and, in the end, impoverished farmers join the ranks of the hungry and indebted farmers commit suicide, as they have in greater and greater numbers in India. Poverty is revealing itself in farmer suicides and the emergence of hunger for the first time in independent India.

India has not had a famine since 1942, but now region after region is experiencing deaths from hunger-starvation deaths. A 1991 government study of a region where 8,000 children died of hunger found, that before trade liberalization and globalization no child in the zero to six age group had died as a result of lack of food.[31] In 2002, 47 percent of children's deaths in India were caused by a lack of food. It is not that there is no food to eat—65 million tons are rotting in the godowns (storage containers). Having disturbed both ends of the balance—in food production and consumption—we now have a world where the grain giants take our food at half the price that the poor pay for it and dump it on someone else's market. Simultaneously, they import food from somewhere else with a $400 billion subsidy that goes not to farmers of the world but to a handful of corporations and dump it on India's market. The promises of peace and prosperity are totally elusive.

The WTO's Agreement on Agriculture, which paved the way for the imposition of cash crops, should be called the Cargill Agreement. It was former Cargill vice president Dan Amstutz who drafted the original text of the agreement during the Uruguay Round. WTO rules are not just about trade. They determine how food is produced, who controls food production. The primary aim of Cargill, and hence the Agreement on Agriculture, is to open Southern markets and convert peasant agriculture to corporate agriculture. But opening markets for Cargill implies the closure of livelihoods for farmers. Asia happens to be the largest agriculture economy of the world, with the majority of the population involved in agriculture. For Cargill, capturing Asian markets is the key. Converting self-sufficient food economies into food-dependent economies is the Cargill vision and the WTO strategy.

Because the Agreement on Agriculture is an agribusiness treaty, it is a distorted vision of production and trade. It is a recipe for ecological destruction, devastation of family farms, and ruination of people's health. Behind the apparent neutrality of rules for domestic support, market access, and export competition are biased opinions and myths about food production and distribution.

Among the myths Cargill perpetuates and enshrined in the WTO Agreement on Agriculture are the ideas that the US is the best region for growing food and that the US grows the best food. In reality, the US is a model of how not to grow and produce food. In 1990, nearly 22 percent of US farming households had incomes below the official poverty threshold, twice the rate for all US families. In 1993, over 88 percent of the average farm operator's household income was derived from off-farm income. From 1982 to 1993, the costs of inputs bought by farmers increased three-fold, driving farm incomes down, as a result from 1982 to 1992, while 67,000 people per year entered agriculture, 99,000 per year left, resulting in the net loss of 32,000 farmers per year.[32] Is it any wonder that during the period from 1990 to 1994, Indian farmers saw an almost minuscule 1.98 percent return on their investment?

The displacement of small farmers has been justified on grounds of alleged productivity of large farms. In fact, as former Indian prime minister Charan Singh has stated, small farms are more productive than large ones.

> Agriculture being a life process, in actual practise, under given conditions, yields per acre decline as the size of farm increases (in other words, as the application of human labour and supervision per acre decreases). The above results are well-nigh universal: output per acre of investment is higher on small farms than on large farms. Thus, if a crowded, capital-scarce country like India has a choice between a single 100 acre farm and forty 2.5 acre farms, the capital cost to the national economy will be less if the country chooses the small farms.[33]

However, it is the small farms and small farmers who are being destroyed by globalization and trade-driven economic reforms. Five million peasants' livelihoods have disappeared in India since "reforms" were introduced.

Another myth is that free trade allows food to be delivered efficiently. The reality is that without massive subsidies and dumping, US corporations could not capture Southern markets. The free trade of agricultural products is basically a food swap, with countries importing the same commodities they export, rather than exporting what they can uniquely produce and importing what they cannot, the entire world is being pushed into trading a handful of commodities controlled by the agribusiness giants.

Furthermore, the idea that globalized agriculture and dumping will free up farmer's incomes to purchase "motorbikes, cellular phones and computers" is a myth which hides the reality that dumping destroys domestic markets, which in turn destroys livelihoods and incomes.[34]

Displacement of farmers and destruction of soil, water, and biodiversity are two negative dimensions of the US food system.

The threat to public health is another fatal aspect of an industrialized, corporate-controlled food system. As US food culture spreads through globalization, it spreads health hazards. The intense controversy over high pesticide residues in Coke and Pepsi in India is one example of the public health hazards posed by US-style industrial food culture. The epidemic of obesity is another. Nearly 70 percent of children in the US suffer from obesity and exhibit metabolic disorders formerly seen only in adults, such as diabetes, high blood cholesterol, and high blood pressure.[35] Today 44 million American adults are obese and another 6 million are "super obese." Obesity is now second only to smoking as a cause of mortality in the US. The Centers for Disease Control estimates that about 280,000 Americans die every year as a direct result of being overweight.[36] The annual health care costs in the US linked to obesity are $240 billion, with an additional $33 billion spent on diet products and weight loss schemes.

With globalization, this bad food culture dominated by profits has spread worldwide. As McDonald's, Coca-Cola, and Pepsi expand their markets, they destroy healthy, local eating habits. The obesity epidemic of the· US begins to spread globally with industrial, junk food. In China, 30 percent of children in 12 schools were found to be obese. In India, nearly 7.5 percent of all children are obese. In Chennai, 18 percent are overweight. Two in five Delhi students have high cholesterol and diabetes. Besides, the health hazards of industrial foods and junk foods, the US is now becoming a source of new hazards in the form of genetically modified organisms (GMOs). Europeans have refused to consume GM foods. India and Zambia refused GM corn as food aid. There is a global treaty, the Biosafety Protocol, to regulate trade in GMOs. However, the US, driven by the biotech industry and agribusiness, would like trade in GMOs deregulated and citizens denied the freedom to know and choose. The false faces of organic corporations—Odwalla juice being owned by Coca-Cola and Celestial Seasonings tea being owned by Hain whose investors include Phillip Morris, Monsanto, and Exxon-Mobil—testifies to the food giants duplicitous intentions.[37] The US threats to initiate a dispute against the EU over GMOs is an example of how WTO rules enable the imposition of bad food and deny countries and citizens their right to food safety and good food.

Recent results have just come out from England from three-year field trials organized by the Ministry of the Environment, and at least for canola (rapeseed) and beet, the results are showing that there is a five-times-higher level of extinction of species in farms that are using genetically engineered organisms than there is in conventional chemical farms. Such studies indicate that, despite the case launched by the US in May 2003 arguing that the Europeans, by not eating genetically modified food, were preventing the Africans from solving hunger problems, it is not going to be very easy to get the Europeans to remove their de facto ban on GM food.

The Europeans are getting fed up with this bullying. Recently European environmental commissioner Margot Wallstrom said:

> They tried to lie to people, and they tried to force it upon people. It's the wrong approach. You cannot force it upon Europe. So I hope they have learnt a lesson from this, especially when they now try to argue that this will solve the problems of starvation in the world and so on. But come on. . . . it was to solve starvation amongst shareholders, not the developing world.[38]

Another myth is that globalization creates a knowledge society. But we are not living in a knowledge society if we don't have the very basic choices that allow us to lead a human life, a life of dignity; allow us to know how our food is produced, allow us to know what kind of forest our tables and chairs are made from, allows us to know whether the wages of the people who grew the food are just or not, allow us to know what's in our food. That's not a knowledge society. Knowledge is not the manipulated data of Monsanto. Knowledge is informed citizens making free choices. That would be a knowledge society. But that democratic frame work is precisely what corporate globalization tries to annihilate.

In India, there were about 20 independent studies on the failure of genetically engineered cotton. The Research Foundation for Science, Technology, and Ecology (RFSTE) carried out studies, the Agricultural Research Institute carried out studies, the Indian government's Department of Agriculture carried out studies, and all the data agreed because it reflected what was happening in farmers' fields. But the only study you will read in an international setting is an article in *Science* written by Marin Qaim and David Zilberman, two scientists—one from Bonn, one from Berkeley—showing a 80 percent increase in yield. Yields were actually down to 10 percent.[39] These scientists never came to India during the planting season. They were given data by Monsanto and they published it. But this is not exceptional. This is exactly what happened when the recombinant growth hormone was pushed through. The data for the papers published in *Science* by the so-called FDA scientists was not generated by them. They had never looked at the raw data. Monsanto gave them a ready-made paper and it was published under their names.[40]

The independent knowledge, the little bit that exists, comes about by collaboration between public scientists, public intellectuals, and activists. What is happening to knowledge and research, under the influence of corporate science with its monopoly on knowledge, is dangerous for the human condition. We can't afford it, especially when the work in science, appropriated by commerce, is generating new threats to the environment and to health. We need more independent knowledge, not less. And instead we are being misled; knowledge is being censored so that free markets can be created.

Contemporary Enclosures

The enclosures of the commons we witness today, in the privatization of water and patents on life forms and biodiversity, are rooted in the first enclosure movement, which has been called the "revolution of the rich against the poor."[41] The enclosure of biodiversity and knowledge is the latest step in a series of enclosures that began with the rise of colonialism. Land and forests were the first resources to be enclosed and converted from commons to commodities. Later, water resources were enclosed through dams, groundwater mining, and privatization schemes. Now it is the turn of biodiversity and knowledge to be "enclosed" through intellectual property rights (IPRs).

Contemporary enclosures enclose both resources and culture. A recent report from the International Forum on Globalization expands the use of the term *commons* to include public services like health, water systems, education, and information.[42] The management and ownership of commons can be structured in a variety of ways; what matters is ensuring the common good and common interest of all people, not just that of a privileged few. The state can either facilitate enclosures and privatization of the commons, or by creating public systems and social welfare structures, it can uphold the commons that serve the common good. It is this ambivalent role of the state that makes it a zone of contest in conflicts between enclosures and reclamation of the commons.

Intellectual Property Laws

Commons are the collective economic assets of the poor. Enclosures of the commons are thus a theft of the resources on which the poor depend for their livelihood. Human survival in India, even today, is largely dependent on the direct utilization of common natural resources.[43] Ecology movements are voicing their opposition to the destruction of these vital commons so essential for human survival. Without clean water, fertile soil, and crop and plant genetic diversity, economic development will become impossible. Sometimes by omission and sometimes by commission, formal economic development activities have impaired the productivity of common natural resources, which has enhanced the contradiction between the market economy and the sustenance economy. Common property resources involve a combination of rights and responsibilities among users, balancing use and conservation, a sense of partnership with nature, and sharing with diverse communities.

Within indigenous communities, innovation is seen as a social and collective phenomenon and the results are freely available to all. Consequently, not only biodiversity but its utilization has been in the commons and has been freely exchanged both within and between communities. Innovations have been passed on over centuries to new generations and adopted for newer uses. Over time, these innovations have been absorbed into the common pool of knowledge. This common knowledge has contributed immeasurably to the vast agricultural and medicinal plant diversity that exists today. This heritage is not viewed as a good with an owner for the purpose of extracting economic benefits. Heritage is not property at all—it is seen as a bundle of relationships rather than a bundle of economic rights representing community and individual responsibility. The concept of individual "property" rights to either bioresources or knowledge remains alien to the local community. This undoubtedly exacerbates the usurpation of the knowledge of indigenous people, with serious consequences for them and for biodiversity conservation.

When the WTO put in place intellectual property rights and prefixed them with TR meaning "trade related" they intimately linking intellectual property with trade and dramatically transformed the idea of intellectual property. Previously, patent law was determined by individual countries according to the situation in the country and rewarded inventiveness and creativity. Each community decided, according to the social situation of the people, the limits for rewarding creativity—what was the common property of the people and what could, for a short time, be treated as an exclusive right. At no point before 1995 did intellectual property cover the very life forms of this planet. After 1995, when the laws of trade-related intellectual property rights (TRIPs) came into force, not only could cells, genes, plants, sheep, and cows be owned as intellectual property, life *had* to be owned. That's what Article 27.3(b) of the intellectual property agreement of the WTO imposed on the world. The consequences of this legislation are, of course, tremendous. Our relationship with the rest of the living world is no longer that of partner, but one of consumer and, for the corporations, that of creator.

Under the zamindar system the British created landlords to collect rent in order to run the empire. The resultant famine of 1942 which killed 2 million people, and that extraction and appropriation of rent is nothing compared with the annual rent collection that accompanies the transformation of seeds into intellectual property. The WTO has created lifelords and tasked corporations with taxing the peasant who has to save seeds and the AIDS victim in Africa who has to get medicine. A farmer saving seed or exchanging seeds with a neighbor is to be treated as a criminal.

This transformation denies history. The only reason we still have seeds in the world is because people saved seeds. Not only did they save seeds, it was considered unethical not to save seeds. Ancient Sanskrit texts say, "The highest sin is to allow the seed to go extinct." In my region in the Himalayas, during the Gurkha war, there was starvation but in not a single hut were the seeds eaten. The seeds had been left untouched. The people went hungry but the seeds were left for future generations. But now we are not only designing technologies that eliminate the fertility, the reproducibility of the seed (the famous terminator technology), we are implementing legislation on property rights that basically says, "Living things that reproduce should not reproduce." Everything has to be turned into a commodity.

Privatization of Water

During the WTO Ministerial meeting in Doha, a clause was snuck into the declaration which talked about the "removal of all tariff and non-tariff trade barriers" to trade in environmental services. "Trade in environmental services" basically meant trade in water. In Cancun, the proposal called for the "removal of all tariff and non-tariff trade barriers to trade in environmental goods." That was still water. Whether called a good or a service it is still free trade rights in water. Free trade rights in water creates situations like the one which allowed a California company, Sunbelt, to sue Canada under NAFTA because the Canadians said, "No, we don't want to sell our water."

These treaties are about more than goods moving across national boundaries. They are about commoditization of the entire planet and transformation of the very basis of life-the planet's life and human life—into corporate property. WTO rules enclose the commons and deliver them into the hands of just a few companies. In water there are five water giants: Suez, Vivendi, Bechtel, Thames, and RWE. And Halliburton has just entered the field in oil-rich northeast India. Bechtel you may know as one of the first companies to get big reconstruction contracts in Iraq. Bechtel was also the company that tried to take over the water contract in Cochabamba, Bolivia, but was thrown out by the collective action of the people. Under Bechtel's interpretation, the water concession meant every drop of water in the region was their property and a rural woman using a pail to draw water from her own well was a thief. In the US, Bechtel is profiting from the currently $14.6 billion highway construction project called the Big Dig. Here, too, water is evading Bechtel's control as the tunnel is "riddled with leaks."[44] As with the first enclosures, privatization serves to benefit a few at the expense of the many.

The argument behind the privatization of water is that because some investment of work is made, or some corporate capital is invested, the resource itself should be redefined as private property. Advocates for the privatization of water argue, "Purifying water from the raw state, treating it, bringing it to people, taking it away again, is so much work. This is what renders it an industrial product." But these same people fail to recognize the work nature contributes by taking water down from the mountains, transporting it thousands of miles to the sea, evaporating it, and returning it back to the earth.

Recategorizing water as private property creates the possibility of excluding others from access to that which is necessary for living. A system which can claim that water will be allocated on this planet according to how capital can control and access it is saying that most species can go extinct. No species seeks its entitlement to its share of water through the market place; they get their access to water through being members of communities and ecosystems.

The eucalyptus, which is so beautiful in Australia, has created havoc in India where it robs other species of their share of water. The eucalyptus was introduced inappropriately, as a monoculture cash crop, and acts as an alien species in an environment where it doesn't fit. Trade treaties and the commoditization of water act in the same way.

In water privatization, as in the other fields where the privatization of life is directly involved—biodiversity and food—not a single project is actually fulfilling any of the promises to better the human condition. Privatization works for corporations even when it fails. For the public it is a recipe for losing access to public services while being trapped in debt. In Manila, the water was privatized but Suez could not run the system. It pulled out, leaving the public utility with the responsibility of supplying the water, and burdened with debt to the World Bank and guaranteed payments to Suez. In South Africa, 10 million people had their water service cut off because they couldn't pay. As a result 300,000 got cholera and 300 died. In India—as elsewhere where people argue, "we need to have privatization because without this money, our systems won't work"—what you find is that it is public money that makes privatization run, and at 10 times the cost of operating the system publicly. That's what happened with Delhi's privatization which has led to a 10 fold increase in water tariffs while all investments are made with public funds.[45] Whether it's a university or a health care system or a water delivery system, that same system with public money could have been run at one-tenth of the cost. Privatization is proving to be extremely ineffective and inefficient for the public. It is proving to be highly effective for the corporations who not only dismantle the public domain but walk off with guaranteed incomes long after they have failed to deliver health or water or energy.

Enron, for example, totally failed in India but is still claiming payments. Bechtel, General Electric, and Enron are claiming $1.2 billion for a project they could not complete and for energy priced so high that they could not sell it. And yet whether or not they provide energy, they are guaranteed payment for 20 years. All privatization contracts come with these guarantees. And they call it the free market. The market does not provide services. The market locks in public resources and public wealth to generate incomes and profits for corporations.

"Takings" and Enclosures

The enclosures and recovery of the commons is not just a historical issue for England, or an issue at the heart of political and economic conflict in contemporary India. It is at the core of political conflicts in the US and at the center of debates on globalization worldwide.

The dismantling of the public domain of common security and public good, carved out through legislation related to environmental protection, protection of labor rights, social security, public health, and public education, is a form of contemporary enclosures of the commons and it rests on past enclosures. The controversy surrounding the appointment of judges to the US Supreme Court is, in effect, about attempts to enclose the commons recovered through the public trust doctrine.

The first enclosure in the "New World" took place when the land and resources of indigenous communities, the original inhabitants of the Americas, were taken over through violence and the worst genocide in human history.

Religion, a mechanistic world view, racism, and cowboy capitalism combined to appropriate and take land and territory that belonged to the original inhabitants. The economy based on enclosures was the economy of the robber barons, the merchant adventurers and pirates, the cowboy colonizers. The colonizers and occupiers then assigned to themselves "natural rights" to property, as articulated by the philosopher John Locke. The creation of private property through enclosures of the commons was defined on the basis of removing resources from nature and mixing them with labor. The labor of indigenous cultures was made part of nature; land was thus made empty of human contribution. Terra Madre was transformed into terra nullius, an empty earth to be carved out as private property by cowboy capitalists.

This violent economy of dispossession and unregulated capitalism was tamed by the New Deal, and some commons were recovered through the public trust doctrine. National parks, beaches, and waterways were protected as commons, with the state as trustee.

The attack on environmental legislation in the US is an attempt to undo the recovery of the commons through the New Deal in order to enclose all public goods and resources as private property. Furthermore, this theory seeks to simultaneously define such enclosures and takings as a natural right while defining the recovery of the commons as "takings" and theft.

Richard Epstein's book, *Takings: Private Property and the Power of the Eminent Domain*, is the bible of the "ownership society" of the cowboy capitalists of the 21st century. It is also the bible of judges like Clarence Thomas and Antonio Scalia who have used Epstein's philosophy of takings to undo the Clean Water Act, the Endangered Species Act, and alter laws based on the public trust doctrine. The problem with Locke and Epstein is that they are blind to the takings by colonizers, cowboys, and corporations, they elevate property created through theft of the commons into a sacred category, defining all attempts to protect the common good as a taking for which the original "takee" must be compensated. As Epstein says in *Takings*, regarding government's environmental protection regulations to prevent enclosures of beaches and pollution of streams, the government must "provide benefits to the individuals who have been coerced that leave them at least as well off as they were before the coercion took place."[46]

If this logic was applied fully and honestly, the Native Americans should be getting compensated for the coercive takings of their land. However, the takings from local communities are erased, and when the government tries to protect beaches and streams as public goods, it is defined as a taking. Thus in the case of *Nollan vs. California Coastal Communities 1987,* when the government tried to protect the beach for the public's common access, Justice Scalia ruled that it was a "taking"—unjust and unconscionable. In *Lucas,* a similar case, Scalia gave a similar ruling. In *Dolan,* restrictions on construction to prevent flooding were also defined as taking; hence illegal.[47]

The entire foundation of the "ownership society" is based on new enclosures. And the contrived law to justify contemporary enclosures *à la* Epstein is based on three falsifications.

The first is the erasure of the history of colonization as a taking, and the denial of the experience of occupied inhabitants and their prior rights and prior claims.

The second is the defining of the behavior of states acting on the public trust doctrine as the same as the eminent domain. Public trust recognizes community rights of people to common property; common goods, and community resources as the highest rights, with government delegated as the trustee to protect the common wealth. The eminent domain principle is based on denial of community sovereignty, and enables government to act against the public—and the common good.

Governments protecting forests, beaches, rivers, atmosphere as commons are acting on the public trust doctrine. Governments enclosing commons and displacing people act on the eminent domain as in the case of displacing people for dams, highways, and shopping malls. The public good is sacrificed for private gain, though it is always the public interest that is invoked.

The third deliberate distortion is the reduction of *public* to *individual.* Public is used both for government as well as collective interests and community organizations. However, cowboy capitalism reduces society to individuals, and makes community disappear. Margaret Thatcher said there is no such thing as a society, there are only individuals. Ayn Rand has said there is no such entity as the public, since the public is merely a number of individuals.

Richard Epstein argues takings is also based on making community welfare and the public disappear. He claims the public interest is the sum of all private interests, and privatization of public goods is a system designed to advance community welfare. He is able to push this false argument on fabricated data. According to him, environmental laws allow the public to gain $1, but the private property owner loses $10;

therefore, since the public interest is the sum of private interests, the public loses through environmental legislation. In contrast, our study on costs of shrimp farming showed that for every $1 of profits for the shrimp industry, $10 of losses were borne by the local communities. Without the Coastal Regulation Zone law for protecting fragile ecosystems, the Indian Supreme Court could not have ordered the closure of the shrimp farms. That is why in India, as in the US, corporations which extract superprofits by destroying nature's economies and local sustenance economies are trying to dismantle environmental legislation. They hide their brutal takings from nature and society by declaring laws that protect nature and society as takings by government.

This is the agenda for privatization and enclosures through deregulation for the cowboy capitalist.

Loss of Workers' Security

For the commoners and the community, enclosure creates new poverty and new insecurity. Instead of land, biodiversity, and water being the source of livelihoods and economic security for the poor, their labor is the only "resource" left to· them. The rise of capitalism replaced producer-run economies with capital-run economies. In preindustrial England, craft guilds ensured a fair living to its members and a high quality of craftsmanship. Mutual help in sickness and poverty was an essential part of the guild system.

> The essence of the guild system lay in the control of industry by the industrial workers themselves, through an elected authority appointed by them. In the capitalist system on the other hand, this control is transferred to men who stand outside the ranks of the industrial workers, and are frequently in conflict with them.[48]

Through years of organizing and pressure in the struggles for workers' rights, jobs were made secure in the industrialized countries and the "organized" sector in the South. Today, as a result of globalization, workers' rights are being dismantled and instead of growth generating employment, we are witnessing jobless growth. According to Jeremy Rifkin, corporations are eliminating more than 2 million jobs annually in the US.[49]

One of the victories workers have won is the promise of corporations to provide for their security in their old age. Corporate globalization is helping companies to avoid these obligations. By outsourcing jobs, corporations escape from their obligations to their employees in the pursuit of maximizing their profit margins. According to the consulting firm McKinsey, because of global pay gaps, stemming from, among other causes, a benefit gap, corporations save at least 45 to 55 percent of their total costs by outsourcing. The wage gap between US and India is 12 to 1 for telephone operators. According to a 2003 study by the University of California at Berkeley, corporations could reduce their expenses by around $300 billion a year by outsourcing an estimated 14 million US service jobs.[50]

Corporations also escape their Social Security obligations by the outright violation of contracts. Companies have gone so far as to, as the November 10, 2004 headline of the *Wall Street Journal* puts it, "Sue Union Retirees to Cut Promised Health Benefits." The article reported on a court summons received by retiree George Kneifel when Rexam Inc., a beverage can maker, sought to eliminate contracted health care benefits.

Furthermore, the proposed privatization of Social Security—high on President Bush's agenda for his second term—gives corporations "freedom" to exploit young workers and reclassifies elders as a burden on society, a new category of disposable people. In fact, these elders have given their share to society in their working years and need care in their old age, from families and community, from their employers, and from the government.

The Blindness of the Market

The organizing principles of development based on economic growth render valueless all resources and resource processes that are not priced in the market and are not inputs to commodity production. This premise very often generates economic development programs that divert or destroy the resource base for survival. While the diversion of resources—such as the diversion of land from multipurpose community forests to monoculture plantations of industrial tree species, or the diversion of water from staple food crop production and drinking water needs to cash crop production—are frequently proposed as programs for economic development in the context of the market economy, they create economic underdevelopment in the economies of nature and sustenance. Earth Democracy movements are aimed at opposing these threats to survival from market-based economic development. In the Third World, ecology movements are not a luxury of the rich; they are a survival imperative for the majority of people whose life is put at risk by the market economy and threatened by its expansion.

The market economy views conflicts over natural resources and ecological destruction as distinct from the economic crisis and proposes its own expansion as the solution to the ecological and social crisis it has engendered. Instead of programs of gradual ecological regeneration of nature to restore the sustenance economy, immediate and enhanced exploitation, with higher capital investment, of natural resources is prescribed as a solution. Privatization of water and water markets are offered as a solution to water depletion and pollution which are all "externalities" of the market and which have created the water crisis as I discussed in *Water Wars*. The commodification of biodiversity through patents on life is offered as a solution to the crisis of species extinction driven by the monocultures on which global markets are based. The disease is thus offered as the cure.

Sustainability

We share this planet, our home, with millions of species. Justice and sustainability both demand that we do not use more resources than we need. Restraint in resource use and living within nature's limits are preconditions for social justice. The commons are where justice and sustainability converge, where ecology and equity meet. The survival of pastures and forests as community property, or of a common good like a stable ecosystem, is only possible with social organizations with checks and controls on the use of resources built into their principles. The breakdown of a community, with the associated erosion of concepts of joint ownership and responsibility, can trigger the degradation of common resources.

In each age of enclosures and displacement, progress is invoked to sell a project in which the elite usurp the resources and livelihoods of the poor as the inevitable next step in human evolution. A trajectory of exclusion is presented as improving the lives of the underprivileged, even though upheaval and displacement is often the result. By universalizing the measure of progress and development, enclosures are hidden, and benefits accruing to the powerful are falsely represented as benefiting the displaced and the dispensable. Development as dispensability is thus sold as development as well-being and welfare. This is what happened when dams displaced people. This is what is happening as highways and river diversions—the infrastructure of globalization—are rendering people dispensable. "India Shining" was the slogan used in the massive advertising campaigns of 2004 for projects supporting globalization even though ecologically, economically, and culturally the result is "India Uprooted."

Stability

Sustainable societies move in a stable state—with, not against, the cycles of life. To be in a stable state is not to be motionless; it involves movement and progression within an orbit, like an electron around the atom or the moon around the earth. The ecological consciousness of ancient civilizations allowed them to progress in an ecologically stable way. But just as classical physics is incapable of explaining or understanding the motion of the electron, conventional market economics interpret stability as stagnation and not as movement at all. Indigenous cultures of the Amazon, of the Andes, or the Himalayas are examples of living cultures that have been sustainable over millennia and, where not destroyed by the globalized economy, are sustainable even today. Capturing this conflict, Gandhi stated that modern civilization:

> Seeks to increase bodily comforts, and it fails miserably even in doing so.... This civilization is such that one has only to be patient and it will be self-destroyed.... there is no end to the victims destroyed in the fire of [this] civilization. Its deadly effect is that people come under its scorching flames believing it to be all good.
>
> It is a charge against India that her people are so uncivilized, ignorant, and stolid, that it is not possible to induce them to adopt any changes: It is a charge really against our strength. What we have tested and found true on the anvil of experience, we dare not change. Many thrust their advice upon India, but she remains steady. This is her beauty; it is the sheet anchor of our hope.[51]

Contemporary ecology movements represent a renewed attempt to establish that steadiness and stability are not stagnation, and that balance with nature's essential ecological processes is not scientific and technological backwardness, but rather a sophistication toward which the world must strive if planet earth and her children are to survive. At a time when a quarter of the world's population is threatened with starvation due to the erosion of soil fertility, water, and genetic diversity, chasing the mirage of unending growth becomes a major source of genocide. Killing people through the destruction of nature is an invisible form of violence which threatens justice, peace, and survival. Editor and author Claude Alvares calls this destruction the Third World War, "a war waged in peacetime, without comparison but involving the largest number of deaths and the largest number of soldiers without uniform."[52]

In a stable constellation of economic organization, nature's economy is recognized as the most basic, because it provides the foundation for the sustenance and market economies, and because it has the highest priority to and claim to natural resources. However, forces for development and economic growth treat the market economy as primary and nature's economy and the sustenance economy as marginal and secondary. Capital accumulation does lead to financial growth, but it erodes the natural resource base of all three economies. The result is a high level of ecological instability, as illustrated in the ecological crisis created by commercial forestry, commercial irrigation, and commercial fishing. To resolve ecological conflicts and regenerate nature, these economies must be given their due place in the stable foundation of a healthy nature. The stability of an economy which values nature as opposed to one which values capital is demonstrated in FIGURE I.

Figure I

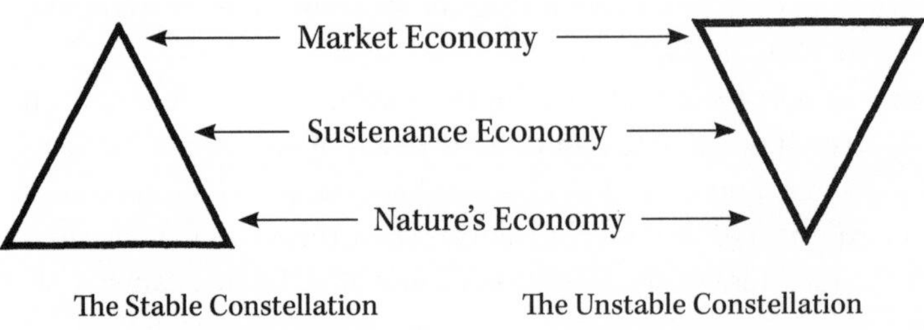

Development and economic growth are perceived exclusively in terms of processes of capital accumulation. However, the growth of financial resources at the level of the market economy often takes place by diverting natural resources from people's survival economy and nature's economy. This generates conflicts over natural resources, and it also creates an ecologically, socially, and politically unstable constellation of nature, people, and capital.

The insatiable appetite of growth and the ideology of development based on it are the prime factors underlying the ecological crises and the destruction of natural resources. The introduction of unsustainable cash crops in large parts of Africa is among the main reasons for the ecological disaster in that continent. The destruction of the ecological balance of the rainforests of South America is the result of the growth of agribusiness and cattle ranching in the clear-cut areas. And with no obligation to ecologically rehabilitate the ravaged land, agribusiness just moves on to consume other resources and other sectors to maintain and increase profits when the productivity of the land declines. The costs of the destruction of Africa's grazing lands and farmlands, and of Latin America's forests, have not been borne by multinational food corporations but by the local peasants and tribals. The costs of ecological destruction and damage to the sustenance economy are borne by the local populace alone.

The False Tragedy of the Commons

The capital-intensive machines and processes of industrialization have become the measure of human progress, and this progress has been used as a justification for enclosures and privatization from the first industrial revolution to the current biotechnology revolution. From the perspective of the powerful, the enclosure of the commons leads to progress, development, and growth. From the perspective of the people, the enclosures create new poverty, powerlessness, and, in the extreme, disposability.

Enclosures polarize the common interest of people into the interest of the rich and powerful and the poor and marginalized. Granting the right to privatize the commons and create private property through enclosures goes hand in hand with pushing millions into deprivation. The story of the commons will always be distorted by those who seek their enclosures. Greed, domination, exclusion are not "essential" human qualities. Functioning commons demonstrate that people can govern themselves, that democratic self-organization and self-governance work, and that people can cooperate, share, and jointly make democratic decisions for the common good. Those who want greed rewarded through the private takeover of common property deny cooperation, sharing, and self-regulation can work in human society.

The philosopher of enclosure, Thomas Hobbes, viewed life as a "short, nasty, brutish affair" and argued that man is constantly engaged in a war of all against all. This view of humanity as inherently competitive denies the sustainability of the common.

In the 20th century, Garrett Hardin revived the tragedy of the commons and turned it into a science. Hardin set up a scenario where each user of the common is faced with a choice that will bring a large individual gain at a small collective cost—adding a cow to one's herd versus the impact on the common. He argued that a common pasture would inevitably deteriorate since every herdsman will act in their own self-interest and put more and more cattle on the pasture. This is Hardin's tragedy of the commons.

What Hardin did not see is that the very existence of the commons implies the reality of cooperative management and ownership. It is important to recognize that competition has not always been a driving force in human societies. The scientist and philosopher Peter Kropotkin writes:

> If we . . . ask Nature: "who are the fittest: those who are continually at war with each other, or those who support one another?" we at once see that those animals which acquire habits of mutual aid are undoubtedly the fittest. They have more chances to survive, and they attain, in their respective classes, the highest development of intelligence and bodily organization.[53]

In large sections of rural societies of the Third World, the principle of cooperation still dominates. The poor could not survive if they did not participate in economies of cooperation and mutuality. Similarly, production for one's own consumption rather than for exchange has long been the predominant mode in subsistence economies. In a social organization based on cooperation among members and production based on need, the logic of gain is entirely different from that of societies based on competition and profits. The general logic underlying Hardin's tragedy of the com mons does not operate under such conditions.

Hardin also failed to recognize that the community jointly decides how many cattle will be allowed to graze on the common and in which season. By the very nature of the commons, the poorest cannot be excluded nor, as long as collective community management is in place can it be privatized by the powerful. A privatized commons is no longer a commons, it is private property, either de facto or de jure. What has been called the tragedy of the commons is, in fact, the tragedy of privatization. The degradation Hardin projected on the commons results from the ability of the powerful to exploit resources beyond the ecological limits of renewability. Daniel Fife, a scientist and researcher, points out:

> The tragedy of the commons may appear to be occurring but in fact something quite different is really happening. The commons is being killed but someone is getting rich. The goose that lays golden eggs is being killed for profit.
>
> That situation is all too possible in the business world. Responsible business ensures that it can continue to run indefinitely. But when a business adopts "higher temporary profits" as its principal goal, its irresponsibility may lead to the destruction of its own resources.[54]

The erosion of systems of social control resulting from modernization and development has led to Hardin's model of degradation of the commons in most regions. However, under certain circumstances where common lands cannot even support the basic needs of the population, a tragedy is to be expected even in the absence of competition.

Village commons have been a historical reality in India. Relics of village woodlots or roadside plantations can still be easily found. In the traditional village, private and unequal landholdings existed side by side with common and equally shared resources. While self-interest might guide a landlord's use of his own land, there were controls, even for the private landlord, over the use of common resources.

Social and cultural regulations have been the main mechanisms for preventing the exploitation of nature. Communities are based on commonly accepted norms and values, which provide the organizing principles and control mechanisms for its members. A shared resource can be managed communally through the implicit acceptance on the part of all the members of the community of a commonly shared norm.

The self-sufficient nature of the traditional village economy maintained the commons despite socio-economic inequalities. Self-sufficiency provided a level of equality which prevented individuals from undermining community action. Thus, for example, in a traditional coastal fishing village with its own socio-economic hierarchies, the exploitation of common resources (the fish in the ocean) was guided by rigid controls to which everyone was subjected. The exploitation of the poorer sections of the village took place on the shore when the catch was distributed on the basis of private ownership. However, the most powerful groups were prevented from overexploiting the resources of the sea by community regulations. Examples of regulation could include not fishing during spawning seasons or not using fishing nets with a fine mesh. Such policies are the primary reason why India's marine ecosystem was maintained over the centuries. The conservation of village woodlots was guaranteed through similar mechanisms. It was when the simultaneous adherence to individual and community regulations was no longer essential, because of the intrusion of large urban and industrial markets into the village economy, that community regulation of common property became threatened.

Access to the bigger markets was, and still is, by and large, possible only for the most privileged members of the community. The easy access to educational, bureaucratic, and financial institutions initiated a process whereby the rich were no longer subject to traditional social norms; this, in turn, led to the breakdown of the community. In the case of marine resources, the introduction of mechanized trawlers (through international and local funding) led to the violation of traditional community norms and influenced the manner in which marine resources were exploited. Similarly, the introduction of new agricultural techniques that were adopted only by the rich farmers made the village elite less dependent on local resources (for example, internationally manufactured chemical fertilizer in place of locally produced green manure). Such circumstances led ultimately to the slow decay of local resources and the community norms which had governed their use.

Myths About and Reasons for Overpopulation

There is a growing literature which seeks to blame people in the Third World and, in particular, their population growth for our current ecological crisis. A prime advocate of focusing on the poor in order to address the environmental crisis of the rich is Garrett Hardin. In his "Lifeboat Ethics: The Case Against Helping the Poor" the poor, the weak are considered a surplus population, putting an unnecessary burden on the planet's resources. Rather than a philosophy of "women and children first,"

Hardin's lifeboat operates on a triage basis—and in times of crisis the weak get sacrificed.[55] In this paradigm of security, the control over and demand for resource of the powerful is what needs to be protected. This was the message given by the US administration when then president George H. W. Bush proclaimed at the 1992 Earth Summit and, nine years later, shortly after 9/11, Dick Cheney echoed, "The American way of life is nonnegotiable." If resource-destructive life styles are to be protected, some people become expendable.

According to US policy, population control activities are a security issue. This is illustrated in a summary of the Department of Defense's position on population:

> As difficult and uncertain as the task may be, policy makers and strategic planners in this country have little choice in the coming decades but to pay serious attention to population trends, their causes and their effects. . . . They must employ all the instruments of statecraft at their disposal (development assistance and population planning) every bit as much as new weapon systems.[56]

A National Security Policy reports finds growing populations will create growing domestic needs. As a result, "concessions to foreign companies are likely to be expropriated or subjected to arbitrary intervention. Whether through government action, labor conflicts, sabotage, or civil disturbance, the smooth flow of needed materials will be jeopardized."[57] The demands of the US economy give the US enhanced interest in the political, economic, and social stability of supplier countries. Because, as the Center for Strategic and International Studies report finds, manpower-intensive regional conflicts are likely to predominate in the years ahead, the US government has focused on the environment, population growth, and women's rights as driving forces for its foreign policy in a new global politics. Third World populations need to be controlled to ensure natural resources for the growth of US corporations.

This imperialistic view of the relationship between resources and population growth does not perceive that population growth is triggered by the appropriation of resources from the common people. Such appropriation also helps create an environment in which resistance movements can grow, as the paired implementation of NAFTA and the Zapatista uprising in Mexico on New Year's Day 1994 highlights.

Controlling the populations without controlling production and consumption patterns does not address the environmental crisis.

The largest pressure on resources does not come from the large numbers of the poor, but from the wasteful production systems, long distance trade, and overconsumption in the First World. The proposed solution—blaming the victims and failing to address the role economic insecurity and the denial of rights to survival plays in population growth—exacerbates the problem.

Reducing populations of local communities and ignoring the burden of the global market cannot protect the ecosystems of the South. Moreover, most analyses of the relationship between population and the environment ignore the nonlocal demand for resources. It assumes that local population pressure is the only environmental pressure on ecosystems. The "carrying capacity" in the case of human societies is not merely a biological function of local population size and local biological support systems. It is a more complex relationship that relates populations in the North to populations and ecosystems in the South.

Northern demands on Third World resources effectively lower the threshold of resources available to support local populations. In other words, what would be a "sustainable" population size on the basis of the local production, consumption, and lifestyle patterns is rendered nonsustainable by nonlocal demands.

Most ecosystems in the Third World carry not only local populations, they also carry, by satisfying the demands for industrial raw material, the North. For example, the Netherlands uses seven times its land area to meet its natural-resource demands. The concept of an "ecological footprint" provides a way of assessing the nonlocal ecological impact of production and consumption patterns. As Wackernagal and Rees of the task force on Planning Healthy and Sustainable Communities write, "The ecological footprint is a measure of the 'load' imposed by a given population on nature. It represents the land area necessary to sustain current levels of resource consumption and waste discharge by that population."[58]

Amory Lovins, an energy expert, uses the metaphor of energy "slaves" to quantify demands on natural resources, "In terms of workforce," he writes, "the population of the earth is not 4 billion but about 200 billion, the important point being that about 98 percent of them do not eat conventional food." According to Lovins, each person on earth places the equivalent of about fifty people's demands on the earth's resources. Lovin goes on to address the unequal demand on resources in which "the average inhabitant of the USA" uses 250 times the resources as the "'average' Nigerian."[59] A rational environmental approach would direct policy measures to target those who demand the majority of the energy.

False perceptions of the problem lead to false solutions. Even if 80 percent of the world's population—the poor people—were to be exterminated through population control, it would only address a small percent of the environmental problem. The lifeboat will sink anyway because of "population pressure" generated by the rich and their "energy slaves."

Environmental space is the share of resources available to each human given the total available resources and the ability of the earth's ecosystem to deal with pollution. The concept of environmental space allows us to view creating more pollution than the biosphere is able to cope with as a form of enclosure since it robs others of their legitimate share. The US, with 36 percent of the CO_2 pollution and less than 5 percent of the world's population, can be seen as, in effect, "enclosing" the atmosphere.

The theoretical conceptual challenge is to locate nonsustainable use not just in visible local demand but also in the invisible, nonlocal demand for resources. Without this broader vision the drive for "sustainable populations" will become an ideological war against the victims of environmental degradation in the Third World, especially poor women; a war that does not confront the real pressures on the environment, which come from global economic systems and the lifestyles of the rich.

What Hardin and those who propose market solutions for the crisis in ecological and human sustainability fail to notice about the degradation of the commons is that such degradation is accelerated when the commons are enclosed, people are displaced, and resources are exploited for private profit. In England, the enclosure of the commons forced peasants off the land and turned it into pastureland for sheep, making "fat beasts and lean people." The disenfranchised people had no recourse but to sell their labor.

Population growth is not a cause of the environmental crisis but one aspect of it, and both are related to resource alienation and the destruction of livelihoods. A population increase also accompanied the enclosure of the commons in England. The population of England (and Wales) doubled from 7.5 million in 1781 to 16.5 million in 1831.[60] In 1600, the population of India was 'between 100 million and 125 million and remained stable until 1800. Then, dovetailing neatly with the expansion of British rule in India and the reduction of resources, rights, and livelihoods available to the people, the rise began: 130 million in 1845,

175 million in 1855, 194 million in 1867, 255 million in 1871. As Radhakamal Mukherjee concludes, "Pax Britanica stimulated a population increase unprecedented in the preceding centuries."[61]

By 1900, the disparity between population and resources and the overcrowding of agriculture, resulted in unemployment and poverty on a scale unparalleled in any modern civilized community. As Mahmood Mamdani has put it, "Another source of income must be found, and the only solution is, as one tailor told me, 'to have enough children so that there are at least three or four sons in the family.'" When people lose all other kinds of security and the absence of any assured social security, children are the only economic security. Because of inadequate resources and inaccessible or poor health service, an Indian woman has to produce six children to ensure at least one son will survive to take care of her and her husband when she is 60.[62]

After many decades of failed "population control," it might well be more fruitful to directly address the roots of the problem—people's economic insecurity. Giving people rights and access to resources so that they can regain their security and generate sustainable livelihoods is the only solution to environmental destruction and the population growth that accompanies it.

Living Economies

The ecological threats to sustenance demand a paradigm shift. Throughout history societies that have neglected to maintain their sustenance resource base have collapsed after an initial period of growth. The collapse of the Mesopotamian and Roman civilizations, for example, was associated with the collapse of their life-support systems. The threat to the sustenance of the sub-Saharan countries is similarly rooted in the destruction of life-support systems. Societies have never succeeded by following a path of unending growth based on over-exploitation of resources.

Earth Democracy movements are struggles of the disadvantaged and excluded, aimed at conserving nature's balance to preserve their survival. Ecology and justice movements offer the world a future by working to ensure the survival of, and protect the fundamental rights to, the earth's resources. They are movements of marginal communities who have been deprived of the benefits of market- and trade-led economic globalization, but who bear all the costs. They reverse the trend of treating uprooted people as disposable people.

Earth Democracy is a nonviolent response to a war that threatens to destroy us all, even the victors. The practices employed in emerging ecology movements represent incipient attempts at a fundamental restructuring toward justice, sustainability, and Earth Democracy. These movements are small, but they are growing. They are local, but their success lies in their nonlocal impact. They demand only the right to survival, yet with that. minimal demand is associated the right to live in a peaceful and just world. Unless worldviews and lifestyles are restructured ecologically, peace and justice will continue to be violated and, ultimately, the very survival of humanity will be threatened.

Our destinies are out of our control. Earth Democracy is a way to face the real challenge of bringing our destinies back into self-regulation. The principles of Earth Democracy evolved through the convergence of groundwork with communities and the debates over the dominant paradigm. Earth Democracy is about ecological democracies—the democracy of all life. For too many people democracy is periodically voting for leaders who turn their backs and say, "It doesn't matter if you don't want war, I'll still go to war. It doesn't matter if you don't want GMOs. We'll still force-feed you with GMOs. It doesn't matter if you don't want to privatize your education system, we'll still privatize it anyway." This "democracy" does not represent or inspire the people.

Our democracy takes into account whose concerns we must have in mind when shaping our economies and deciding what we do with our food, our water, our biodiversity, and our land. The democracy of all life is a living democracy; it recognizes the intrinsic worth of all species and all people. Because all people and all species are, by their very nature, diverse, it recognizes diversity not just as something to be tolerated but as something to be celebrated as the essential condition of our existence. Without it we are not. And all life, including all human beings, have a natural right to share in nature's wealth, to ensure sustenance—food and water, ecological space, and evolutionary freedom. This is not a right written by states. It is not a right that can be denied by corporations and corporate greed.

Justice and Stability

Gandhi's economic constitution reads:

> According to me, the economic constitution of India and, for the matter of that, the world should be such that no one under it should suffer from want of food and clothing. In other words everybody should be able to get sufficient work to enable him to make the two ends meet. And this ideal can be universally realized only if the means of production of the elementary necessaries of life remain in the control of the masses. These should be freely available to all as God's air and water are or ought to be; they should not be made a vehicle of traffic for the exploitation of others. Their monopolization by any country, nation or groups of persons would be unjust. The neglect of this simple principle is the cause of the destitution that we witness today not only in this unhappy land but in other parts of the world, too.[63]

The suicidal market economy destroys nature's economy and the people's sustenance economy, creating ecologic crisis and economic crisis, while making growth nonsustainable and inequitable. Living economies rejuvenate ecological processes while reactivating people's creativity, solidarity, and interdependence. Robust living economies are people-centered, decentralized, sustainable, and livelihood-generating. They are based on co-ownership and coproduction, on sharing and participation. Living economies are not mere concepts; they exist and continue to emerge in our times. Living economies are being shaped by ordinary people in their everyday lives.

Living economies are based on the vibrant, resilient, and renewable nature's economies and rich, diverse, and sustainable people's economies. Living economies are sustainable and just. They respect the renewable limits of natural resources and share those resources to ensure everyone's needs are met. This is why biodiversity and water must stay in the commons. And why the defense of the commons is the basis of many movements that fall within the constellation of Earth Democracy.

Localization

What happens in Iraq doesn't matter to most people in the US. Except for the families of the soldiers who have been sent there, it's not a pain that most people feel on a daily basis. The distance isolates them. This isolation is why localization, the elevating of local concerns and regulation, is a key tenet of Earth Democracy. Localization provides a test for justice. Localization is a test of sustainability. This is not to say all decisions will be made on a local level. There will of course be decisions and policies made on the national level and the global level, but to reach these other levels they have to constantly pass the screen of living democracy. Authority is delegated to more distant levels of government on the principle of subsidiarity: things are most effectively done at the level closest to where the impact is felt. This principle is an ecological imperative.

Devaluing the role of natural resources—in ecological processes and in people's sustenance economy—and the diverting and destroying of these resources for commodity production and capital accumulation are the main reasons for the ecological crisis and the crisis of survival in the Third World. The solution lies in giving local communities control over local resources so that they have the right, responsibility, and ability to rebuild nature's economy, and through it their own sustainability. This is what living economies are undertaking.

Living economies are based on people's creativity and self-organization. Living economies grow outward, from the individual to the community to the region to the country to the global level. The most intense relationships are at the local level and the thinnest interactions at the international level. That is why living economies are primarily local and decentralized, in contrast to the dominant model, which is global and centralized. Localization and decentralization do not imply isolation or the inability to coordinate with a high level of organization. In living economies small self-organizing systems can network to an extremely complex level of organization. In India, the women of Lijjat Papad and the Tiffin carriers of Mumbai who inspire even the global market giants powerfully demonstrate this.

Regulating the Market

Living economies are grounded by two ecological principles necessary to protect and restore nature and society that free market economists have resisted implementing. These principles are the "precautionary principle" and the "polluter pays principle" as enshrined in Agenda 21 of the UN Conference on Environment and Development (1992), known as the Earth Summit. The precautionary principle calls for not undertaking activities that could cause ecological harm. The polluter pays principle requires that the polluter must pay for any harm done to nature and society and for the costs of the cleanup.

Under the polluter pays principle, those engaged in selling fossil fuels and fossil-fuel-based energy and transportation systems must pay for the impacts of climate change, for processes which reduce CO_2 emissions, and to develop sustainable, renewable energy alternatives. It is now recognized that over the next century the climate will change because of fossil fuel use and greenhouse-gas emissions—especially CO_2. The concentration of greenhouse gases already present in the atmosphere will cause the earth's atmosphere to warm in the range of 1.9 to 5.8 degrees Celsius over the next century. And while the industrialized world accounts for the major share of CO_2 emissions—because changes in temperature and precipitation will have greater impact on the viability of agriculture in the tropics; because people in the South are more dependent on local agriculture; and because sea-level rise will have higher impact on coastal communities and small island states—the South will feel the major impact.

Attempts have been made to draft a climate-change agreement addressing the common problems—but with different responsibilities—stemming from atmospheric pollution. The Kyoto Protocol attempts to put the responsibility for reducing greenhouse gasses on countries. In Earth Democracy the responsibility of resolving the climate change problems would be on the companies—and their CEOs. The responsibility of governments and intergovernmental agreements would be to ensure that production and consumption patterns operate within sustainable cycles.

Living Economies for Rejuvenation of Livelihoods

Livelihoods are the human source of sustenance, meaning, and purpose; they provide a sense of self and of community. Livelihoods are ways of living and means of life. Livelihoods are not "jobs," where one sells labor power to someone else who pays wages. A peasant cultivating food does not have a job, but does have a livelihood. Livelihoods are self-generated. Creative production through "self-employment" needs access to resources and this is what commons ensure. Fishing communities need access to the seas as commons for their livelihoods, forest communities need access to the forest commons for their livelihoods, and farming communities need access to biological and hydrological commons for their livelihoods. The enclosures of the commons rob people not just of their resources, but also of their self-sustainability and livelihoods.

In India, 5 million peasants are displaced annually. Seventy-five percent of India's 1 billion people base their livelihoods on agriculture. In spite of 6 percent annual growth of GNP over the past decade, India's employment in agriculture and manufacturing has declined. Even the services sector is witnessing jobless growth. The much heralded outsourcing of information technology jobs to India currently contributes less than 1 percent of the GDP and employs less than 1 million people—a mere .01 percent of the population. Furthermore, these jobs are restricted to the less than 5 percent of Indians who receive a college education. Thus, agriculture and manufacturing will remain the source of livelihood for most Indians.[64]

While globalization is leading to jobless growth and creating disposable people, living economies ensure work for all. Living economies are based on working for sustenance. They put human beings and nature at the center. In living economies, economics and ecology are not in conflict. They are mutually supportive.

Living Economies in Practice

I was taught my first lessons about the value and worth of nature's economy by the women of Chipko. For the local women, the forests were mothers, providing all needs for sustenance—water, food, fuel, fodder, and medicine. Landslides caused by deforestation were the main trigger of their protests. But logging also made streams disappear and aggravated floods and droughts, which resulted in scarcity of fuel and fodder. Logging in India has been a major source of revenues ever since colonial policies converted the forests from commons managed and used for local needs into timber mines to supply raw material for the empire. The living economy of the earth community faced the killing economy of the market. Throughout the 1970s, in village after village, women would come out and by hugging trees—*chipko* means hug or embrace—prevent the logging companies from destroying their forests. A massive flood in 1978 made the government realize the women were right in saying that the forests were not timber mines, but ecological security. In 1981 the government imposed a ban on logging in the high Himalaya. In the act of embracing trees as their kin, ordinary women mobilized an energy more powerful than the police and the brute strength of the logging interests.

Navdanya: The Living Economy of Food

Navdanya is a network that attempts to shift from the suicidal/genocidal economy of agribusiness imposed by the WTO and the World Bank to living economies of food. More than 200,000 farmers are working to enrich the earth, create prosperity for rural producers, and provide quality food to consumers. Rebuilding nature's economy involves rebuilding soil fertility and the biodiversity of microorganisms. It reintroduces biodiverse farming to both replace chemicals as fertilizers and pesticides and to increase the productivity and nutritional value of crops.

Navdanya farmers are able to reduce their expenses by the 90 percent that was used to buy chemicals and create corporate profits. That saved income finances education and health. The incomes of Navdanya farmers are three times higher than the incomes of chemical farmers—less is wasted on toxins and biodiversity, and fair trade reduces farmer vulnerability to volatile markets and unfair rules of trade.

Navdanya has built living economies at the level of production, processing, and distribution. Biodiverse organic production, combined with fair trade is increasing livelihood security, food and nutritional security, and health. The environment, farmers, and public health are all enriched.

Creating the living economy of food is not separate from creating living democracies or living cultures. Alternatives to corporate dictatorship and food fascism deepen democracy and reclaim cultural spaces. Living food economies are creating the real culture of life as alternatives to the culture of death being spread by global agribusiness and the food industry. Living food economies also challenge the myth of "cheap" food. The cheapness of industrially produced, genetically engineered, globally traded food rests on subsidies for oil, subsidies for chemicals, subsidies for exports, subsidies in the form of tax holidays, subsidies in the form of exploitation of farmers and workers throughout the food chain. If real costs of food were internalized, corporate-controlled food would be too costly for anyone to choose. We need to move from "cheap and nasty" to "fair and healthy."

Lijjat Pappad: A Women's Economy

Communities in India are creating living examples of living economies in the cracks of the market. One such example is Lijjat Papad. Lijjat Papad is a typical Indian snack. Starting from a group of seven women in March 1959 grew a women's organization that now has 40,000 members. These women got together in Gurgaum, Mumbai to roll papads as a source of income. Sales, which totaled only 6,196 rupees in the first year, are today 3 billion rupees. The organization has 63 branches and 40 divisions spread across India. Lijjat Papad's success and connection with living economies lies in its philosophy and organizational structure. The organization's brochure reads:

> The biggest asset of our organization is its philosophy. We do not have the might of Power. We have neither a lot of money nor influential people. Yet our organization has survived. It is running very well in spite of no one in particular running it. In fact, it is growing.
> Many will wonder, how this is possible. How can such an organization run where there is no boss, where nobody passes orders, where all enjoy equal rights? In fact there is nothing to wonder about. The key to success of our organization lies in the treasure of our basic thoughts.[65]

The brochure continues to explain the philosophy behind the organization which has allowed it to grow in sustainable ways while ensuring livelihoods for the women, who work with it. Some key philosophical points include:

> Common ownership—All sister members are its owners. All the profit and loss, whatever it may be, is shared or owned by the members jointly. As their pledge states, "I shall adopt the broader meaning of common ownership.... instead of thinking in terms of I should get more than others, I shall aspire that other should not get less than me."
> Non-discrimination—Our organization belongs to sisters. They may be of any religion or caste, educated or uneducated, rich or poor.
> Voluntarism—Ours is a voluntary organization of sisters, which means that they can willingly join it, if they like the philosophy and practices of the organization and they can leave it of their own free will if, there is any special reason to do so.... No kind of work is considered inferior or superior. For the organization every kind of work is equally important and sisters are free to choose the work they like.

Autonomy and Independence—Our organization does not accept aid or charity. This is our basic philosophy. From the very inception of our organization we have never accepted from anyone charity, donation, or grant and we shall never do it in future.

Ethical business and commitment to quality, not profits—The basic objective of our organization is to make some earning by dint of hard work and live with honor. If we do not make money, the organization will not run, and to make money, it is essential to run the business with great understanding, wisdom, and skill.

Our organization is like a family—Along with the philosophy of running the business wisely, it is also our main objective that sister members of our organization work happily in an organized way with a feeling of mutual trust and friendship.... All sisters are equal. No one is a boss or a subordinate. In respect of status, no one is superior or inferior. They have equal rights in the organization.... It does not matter whether they receive more remuneration or less. That does not make one superior and the other inferior. If there is any difference among them, it is in respect of the responsibilities. Some sisters shoulder less responsibility, others shoulder more. The Sanchalika [branch head; chosen by consensus] of the center has to shoulder the highest responsibility and look after other sisters as the mother looks after the members of the family.

Our organization is a place of worship—We treat our organization as a revered place of worship like a temple, mosque, gurudwara, church.... We believe that society itself is the manifestation of God. So whatever economic activity goes on in the society must aim at bringing about welfare of all including you and me. The economic activity that does not have this purpose and which is done with narrow, selfish attitudes is sure to spell disaster for the society.[66]

Dabbawalas, the Dignity of Labor

Everyday in South Mumbai, more than 3 million people go to work, but many are still able to eat homemade food because 5,000 dabbawalas or tiffin box carriers bring hot lunch boxes from people's homes to their offices. With no documentation, no orders, no bosses, this self-organized network, the Mumbai Tiffin Box Suppliers Association, daily delivers 175,000 lunch boxes. That these deliveries occur during a three-hour period, over 28 miles of public transportation, with only one mistake every 16 million deliveries is astounding.

The dabbawalas are considered the descendants of the soldiers of the great Chatrapathi Shivaji Maharaj, the founder of the Maratha Empire. As Raghunath Megde, the president of the association states, "No one in the association is an employee or employer, all are partners and all are co-owners." They have evolved their own logistics of delivery and their own coding system. Each tiffin box has a number of codes painted on it that identify where the box was picked up, the originating and destination stations, and the address to which it is to be delivered.

The entire network operates on the basis of decentralized units which each consist of 15 to 25 dabbawalas. Each group is independent of the other as far as money is concerned, but they coordinate with each other for the delivery of dabbas. Independence does not imply isolation or fragmentation but interdependence and mutuality. Every dabbawala contributes 10 rupees per month toward the trust and each group manages its money and day-to-day functioning. The network of 5,000 that makes millions of deliveries is self-organized and does not rely on external control mechanisms, externally directed organizational structures, or hierarchical management layers. On the 15th of every month, the association holds a meeting to resolve disputes and address problems.

The network does not just provide self-employment opportunities to its members. By maintaining a link between home and work, kitchen and office, the dabbawalas provide continuity to food cultures and food diversity, and offer a viable working alternative to the fast food monoculture and McDonaldization of food systems in a large metropolis.

These examples of sustainable and self-organized economies reverse the logic of economic globalization, which is based on destroying employment and livelihoods and imposing economic management from either the World Bank and IMF or the WTO trade rules or corporate headquarters.

Conclusion

People-centered economic systems are based on people's creativity, intelligence, and self-organizing activities. Incomes are based on returns on investment of labor and not on returns on capital investment. Ownership is based on work, not capital. Decisionmaking is decentralized. Large networks emerge from interconnectedness of small-scale, self-organized activity. Localization does not imply isolation from the larger world, but self-determination with interdependence.

Living economies, with human creativity at the core, thus mimic nature's diversity, self-organization, and complexity. Every person, every group, every community is its own center, connected to others in mutuality and support. Gandhi captures the structures of living economies thusly:

> Life will not be a pyramid with the apex sustained by the bottom. But it will be an oceanic circle whose center will be the individual always ready to perish for the village, the latter ready to perish for the circle of villages till at last the whole becomes one life composed of individuals, never aggressive in their arrogance, but ever humble, sharing the majesty of the oceanic circle of which they are integral units. Therefore, the outermost circumference will not wield power to crush the inner circle, but will give strength to all within and will derive its own strength from it.[67]

CHAPTER TWO

Living Democracies

Democracy has become a much abused term. Operation Iraqi Freedom promised to bring Iraqis democracy but brought torture, abuse, and the denial of basic needs. Globalization promised to spread democracy under the assumption that free trade equals open markets, which equals open societies. This equation does not hold. The markets of corporate globalization are not open—the trade rules give control to giant corporations. And the resulting societies are not open either. Corporate globalization is creating a dictatorship over food and water, over the most vital aspects of our lives. It is robbing us of our freedoms at the most fundamental level—that of survival. Representative democracy is increasingly inadequate at defending our fundamental freedoms.

Globalization and the Limits of Representative Democracy

Corporate globalization and free trade rules provide immunity to corporations and capital from social and political regulation by citizens and governments. By deregulating commerce, corporate globalization takes the decisions about our everyday lives beyond the influence of democracy to the WTO, the IMF, the World Bank, Wall Street, and corporate boardrooms. It is, in effect, the death of economic democracy.

Corporate globalization destroys local and national economies and the livelihoods and jobs that domestic economies generate in the pursuit of corporate profits and financial growth. This creates insecurity. Insecurity breeds fear and exclusion and provides fertile ground for the emergence of politics based on narrow cultural identities and ideologies of exclusion. Representative democracy in this context becomes increasingly shaped and driven by cultural nationalism. Cultural nationalism emerges as the twin of economic globalization.

Citizens do change governments through the mechanisms of representative democracy. However, the control of corporations and the coercive rules of globalization undermine this change by ensuring that political change does not imply changes in economic policies. No matter which party holds office—the Republicans or Democrats in the US, Labor or the Tories in the UK, Congress or the BJP in India—in reality, corporations rule.

For citizens to reclaim their freedoms, we have to reinvent democracy. We have to deepen it and broaden it. Our idea of democracy is not one where governments are elected to office without the people controlling the power of government or the powers government hands over to corporations. Living democracy reclaims our decision-making powers and capacities. It requires self-organization and self-rule (Gandhi's *swaraj*). We must broaden democracy to include the excluded—disenfranchised communities, children, prisoners, the elderly, and the diverse species of earth. I call this form of democracy, Earth Democracy. We need Earth Democracy to protect our freedoms, to maintain the earth's life support systems, to ensure justice and sustainability, to end conflict and bring peace.

The Crisis of Democracy in the Age of Globalization

We are witnessing the simultaneous growth of two forces—one of globalization, the other of localization; one driven by global corporations, the other by local communities and grassroots movements; one moving power upward, the other moving it downward. Globalization is, in effect, the end of economic democracy. Its rules of free trade are designed to grant immunity to corporations from influence or regulation by the state.

In a democracy, it is the aspiration and will of ordinary people that should guide economic policy. When economic policy is designed by the WTO or the World Bank and the IMF, and insulated from being influenced by people, we no longer have economic democracy, but economic dictatorship. This is why increasingly there is no difference between the

Congress Party and BJP in India and Democrats and Republicans in the US in terms of their commitment to policies of corporate rule. Elections do not change economic policies under corporate globalization, They merely change parties and heads of state. The real heads of state are global corporations whose interests are put above the interests of citizens, whose rights are put above the rights of humans and other beings, whose profits are made the highest value, higher than life and freedom.

The limit of representative democracy to defend people's freedoms is best demonstrated in the largest, most vibrant of democracies—India. The 2004 elections rejected the combined power of market fundamentalism and religious fundamentalism. The people's vote was a vote against corporate globalization, referred to in India as economic reform. However, the first thing the new prime minister and officials announced was that "reforms" would not be reversed, merely given a human face. What the people of India were asking for was a human heart for the economy, a heart that cared about the peasants and the poor, the unemployed and the hungry, not the mask of a human face. Economic democracy requires reversing some of the most inhuman aspects of economic globalization, which are securing corporate profits by pushing peasants to suicide, the poor to hunger and thirst, the youth to unemployment.

Victory in Cancun

The outcome of the WTO ministerial in Cancun stands as a victory of democracy over dictatorship, of fairness over injustice, of the South over the North, of the poor over the rich, of people over profits, of life over death.

Cancun, on the turquoise waters and white beaches of the eastern coast of Mexico, was designed to be a holiday resort, not the site of some of the most intense contests and debates of our times. However, from September 10–14, 2003, it was not tourists who filled Cancun's hotels, but government delegations to the WTO. And protesters, not tourists, filled the streets of the hotel zone, marching and trying to cross the barricades which had been put up to keep dissent 10 kilometers from where the fifth ministerial meeting of the WTO was taking place.

As in Seattle, the Cancun meeting failed. Resistance to the WTO's genocidal policies and rules succeeded in breaking down; negotiations designed to expand the WTO into every dimension of our economies and our lives. That WTO rules are not about fair trade but about life and death was made tragically, but heroically, clear by the suicide of Lee Kyung Hae, a Korean farmer, on the first day of the WTO meeting. Lee and other small farmers and peasants from around the world were camping in the grounds of Casa de Culture outside the hotel zone. On the morning of the first day, the farmers led a march to the ministrial meeting. When they reached the barricades put up to keep people out of trade talks, wearing a large sign which read "WTO kills farmers," Lee Kyung Hae climbed the barricade and stabbed himself He had with him a note reading, "I am taking my life so others may live."

This was not the first time Lee had tried to gain the ear of the WTO. In an article in the April 2003 issue of *Korea Agrofood* Lee wrote:

> Soon after the Uruguay Round Agreement was settled, we, Korean fellow farmers and myself, realized that our destinies are out of our hands already.... I am crying out my words to you that have been boiled so long time in my body:
> For whom are you negotiating now? For the people or for yourselves?
> Stop taking your WTO negotiations of fallacy logics and of words of diplomatic gestures.
> Exclude agriculture from the WTO system.[1]

A decade before Cancun—before the WTO was formed—farmers from the Korean Federation of Small Farmers, which Lee had served as president, had protested in India demanding agriculture be kept out of "free trade" agreements. In the years since that protest the coercion and fraud of free trade has become evident. Free trade is, in reality, forced trade—forced upon small farmers and poor countries. It is also fraud trade because, while lip service is paid to efficiency and a level playing field, efficient small producers are destroyed by the dumping of highly subsidized products on world markets.

Farmers like Lee were robbed of their right to live by WTO rules which forced Korea to open its rice markets to dumping by US agribusiness giants like Cargill and ConAgra. In 2001, the cost of production of rice was $18.66 per bushel in the US but it was sold internationally at $14.55 per bushel. This dumping—selling products at below the cost of production—is legal under the WTO and resisting dumping has been made illegal. Farm prices are in a free fall, driven downward by export subsidies that create unfair and unjust trade, the forced removal of impact restrictions (QRs), and the lowering of tariffs. Even before Cancun, Pascal Lamy, while Trade Commissioner of the EU, announced that the EU would not cut export subsidies. The US had announced that *it* would not cut domestic support. In fact, both the US and the EU have increased farm subsidies since the WTO agreements came into force, despite promises in Marrakech to reduce Northern agricultural subsidies and create a level playing field. The WTO has legalized the increase in subsidies through the creation of blue and green boxes—WTO constructions which allow subsidies to grow. Green box subsidies are related to environmental programs and blue box subsidies are those defined as non-trade distorting because they are linked to production; however, they do distort trade because they distort price. Through this mechanism the explicit subsidies for cereals in the EU decreased by 60 percent from 2.2 billion euros in 1992 to 883 million euros in 1999, but direct payment increased from 0 to 12.8 billion euros. By including export refunds, total cereal subsidies reached 2.985 billion euros in 1999. This is, in reality, an increase of 785 million euros in subsidies from 1992.[2]

US subsidies for cotton production and exports have increased to almost $4 billion after the US Farm Act of 2002. In 2001, the cost of production of cotton in the US was $0.93 per bushel, while the export price was $0.40 a bushel, a dumping of 57 percent. This has increased from 17 percent in 1995. The combination of US agricultural policy that provides subsidies for dumping with WTO rules that force countries to remove import restrictions while preventing poor countries from protecting themselves from the devastating impacts of dumping are leading to the destruction of Third World agriculture. The US farm bill has increased subsidies by $82 billion. It allows the US government to pay cotton farmers the difference between the world market price, $1.23 per kilo, and a fantasy ideal price of $1.57 per kilo. As a result, US cotton farmers received $3.9 billion, most of it going to the giant corporate farmers. With these subsidies, the US has doubled cotton exports and destroyed the livelihoods and incomes of 250 million African cotton farmers since the WTO came into being in 1995.

This legalized fraud in the name of free trade in agriculture was the main reason for the collapse of WTO talks in Cancun. While farmer Lee's martyrdom sent a strong message of resistance from the barricades, a rebellion was also brewing inside the convention center.

Before meeting in Cancun, the US and the EU had already reached an agreement on agriculture that would, in effect, force the South to further dismantle trade barriers, while refusing to reduce export subsidies to US and European agribusiness. A group of 21 developing countries made a counterproposal, insisting on the removal of export subsidies that are killing Third World farmers before further reductions of tariffs. There was a deadlock over the two texts. When the WTO issued a draft declaration on the third day of the ministerial, it failed to reflect any of the concerns of the South. Worse, the demand of African cotton-producing countries for protection from US dumping was trivialized to a paragraph suggesting that Africans abandon cotton production. After the draft declaration was released, a representative from the African nations declared, "If Africans leave Cancun without practical results, they may not return, because so much efforts have led to so little."

The walk-out by WTO members led by African countries explicitly rejected the imposition of new issues such as investment, government procurement, and competition policy and trade facilitation.

But the failure of Cancun, following the failure of Seattle, also points to the need to remove from free trade treaties issues such as agriculture and intellectual property, which are better left to national systems, and better handled as basic needs and livelihood issues than as trade and commerce issues alone.

What the US and the EU wanted to secure in Cancun was the right to continue to dump and to continue unfair trade by supporting their agribusiness interests in use of the WTO's market-access rules to take over world markets. Now that the bullying has been challenged by over a decade of persistent organizing by citizen groups and the new alliances among developing countries, EU trade commissioner Lamy calls the WTO "a medieval institution" and Robert Zoellick, the US trade representative, denigrates delegates from the Third World as the "won't do" group. The rich countries have sent a signal that they will not reform and they will not let the WTO be reformed. It is now imperative to stop the one-sided liberalization that is destroying our farmers and agriculture. It is time to put our national and domestic interests above the greed and deceit of the powerful corporations of powerful countries. It is time to bring back import restrictions as the Indian People's Campaign against the WTO has demanded. The paradigm of trade liberalization was dealt a severe blow in Cancun. It is time to give concrete shape to a fair trade paradigm that builds on robust local and national economies. It is time to put people before profits. It is time to put domestic production before international trade.

No new issues, no enlargement of the trade agenda can legitimately be negotiated in Geneva in light of the failure of the WTO in Cancun and absence of agreement at the ministerial level. The loss of the legitimacy of the WTO in Cancun should be turned into an opportunity to reclaim democratic space and shape our economies to serve life, not corporate profits. Democracy won in Cancun. We must now sustain this victory by ensuring that economic and trade decisions do not violate our constitutions and fundamental human rights. We must ensure that global trade driven by corporate profits does not undermine livelihoods and local economies. Economic democracy can only grow upward like a tree, with its roots in local ecosystems, local cultures, and local economies, its trunk supporting strong and vibrant national economies, and its branches nourishing and being nourished by international trade based on principles of sustainability, justice, and fairness. Cancun's failure can be a victory for the alternatives we have all been striving to build to protect the earth and all her people.

In Seattle and Cancun, people and governments of the South asserted their democratic rights and the process of dictatorship was stalled for the moment, but other processes are being used to derail democracy. The most important is the divide-and-rule policy of religious fundamentalism and ideologies of exclusion, which both act as diversions and as dissipaters of people's democratic energies.

Free Market Democracy and Fundamentalism

Free trade treaties like those managed by the WTO and the structural adjustment programs of the World Bank and the IMF are imposed against the popular will of people. Decisions in the WTO and World Bank and the IMF are made undemocratically. The policies these institutions impose transfer the assets of the poor to the rich and to global corporations. Economic democracy, which involves the participation of all people in the decisions about the economy, about ownership of productive assets, especially natural resources, are not advanced by the free market. Economic democracies have space for the creative productive contribution of all people irrespective of class, gender, race, or ethnicity. Unlike the world of globalization, there are no disposable people in economic democracies. It is the indignity of being treated as disposable that pushes people toward religious fundamentalism in order to retrieve a sense of self of meaning, of significance. This is why globalization breeds religious fundamentalism and free markets create terrorism and extremism, not democracy. As Amy Chua notes in *World on Fire:*

> The global spread of markets and democracy is a principal, aggravating cause of group hatred and ethnic violence throughout the non-western world. Because markets and democracy benefit different ethnic groups in such societies, the pursuit of free market democracy produces highly instable and combustible conditions. Markets concentrate enormous wealth in the hands of an "outsider" minority, fomenting ethnic envy and hatred among often chronically poor majorities.[3]

Because globalization erodes equality, justice, and democracy, it fuels a culture of fear, and religious fundamentalism is fueled by fear. No society is immune from this disintegration along cultural lines. It is no surprise that religion emerged as a driving force in India's elections after the trade liberalization/free market reforms were imposed by the World Bank and the IMP in 1991. And it is no surprise that in the 2004 US elections religious and cultural values, not the war in Iraq or the economy, emerged as the deciding factor.

Through the mutation of positive identities to negative identities, globalization is creating and intensifying culture wars, Economic globalization destroys cultures and positive cultural identity by destroying work livelihoods and job security from which most people draw their sense of who they are. Ecological and economic identities are linked to roots in a place and community. When secure occupations and livelihoods are destroyed, the vacuum of the loss of sense of self is filled by the negative identity, an identity of "not the other." Many farmers in the US and elsewhere can no longer draw their identity from the land because debt and foreclosures have driven them off the land.

These uprooted communities are manipulated on the basis of negative identities to create vote banks and a power base. Democracy is diverted from economic democracy by a culture war based on negative identities. A vicious cycle is thus created in which nonsustainable, anti-people economic policies are kept propped up by negative cultures and a negative democracy which is not "of the people, by the people, for the people" but "of the corporations, by the corporations, for the corporations." Politicians must win elections but cannot intervene in economic processes on behalf of the public. Instead of economic justice, they therefore capitalize on cultural and religious identity and insecurity. Without economic democracy, political democracy becomes a force for dividing people, not uniting them. Cultural diversity and pluralism, the very fabric of democracy, is torn apart by the direct assault of the market and manipulation of politicians.

Reinventing Local, Reinventing Global

Corporate globalization is based on contrived rules of trade which invade our autonomous and sovereign spaces—ecological, economic, cultural, social, political, ethical, spiritual. On the one hand, globalization redefines life as commerce and the world as commodities. On the other hand, it limits our global experience to global markets and global institutions. But global can be understood differently. It can refer to our universal values as humans-experienced in our lives, embodied in diverse faiths, enshrined in the UN Declaration of Human Rights. It can refer to humans as one species among many, which both differentiates us from and connects us with other species. We can experience the global belonging to the earth family. And as members of the earth family, our consciousness compels us to respect and protect the sovereignty, integrity, and ecological spaces of other species. This in turn creates the imperative to reduce our ecological footprint on the earth's resources.

Localization internalizes social and ecological costs in our production and consumption systems—not as an afterthought or as an external arrangement of "trade in pollution permits" in which pollution and injustice grow. By prioritizing people and nature above commerce and profits, ecology and equity above trade, citizens above corporations, local democracy above the global market, and people's lived realities in their everyday life above the abstract constructions of corporate capitalism and multiple patriarchies, resources are conserved, livelihoods are protected, justice and sustainability can be ensured.

Localization is based on the interdependence between nature and culture, humans and other species, local and global, micro and macro. Localization treats every place as the center of the world, placing every person, every being as the center of ever widening circles of compassion and care. Neoliberalism treats Washington as the center of the world and renders everything that does not fit this perspective as disposable. The Washington consensus, based on the inability to recognize diverse beings, cultures, and economies, is simultaneously unable to recognize the ecological and social costs of implementing corporate globalization. This is why the promoters of trade liberalization cannot hear the voices of those who scream as they bear the costs and burdens or are being rendered dispensable. Globalization has no self-correcting mechanism. Localization constantly responds to ecological feedback from nature and political and cultural feedback from people.

Local economies are not the only economy, but they are the foundation of fair and nonexploitative international trade. Localization builds sustainable and just economies that grow from the bottom to the top, from the inside out, from the local to the global. They acknowledge that the artificial subsidies that destroy local production and local livelihoods do not actually make nonlocal production cheaper. The myth of cheap food, which is simultaneously destroying the earth, killing farmers, and creating disease, is based on fossil fuel subsidies that allow cheap long-distance transport and lower prices of nonlocal produce; on export subsidies that reduce international prices and promote dumping; and on agribusiness monopolies that allow corporations to enslave and exploit farmers. Cheap food is indeed very costly to the environment, farmers, and public health. It is cheap because it does not reflect real costs. It is a false economy. Localization creates honest economies based on real costs, real prices.

New initiatives such as Community-Supported Agriculture, fair trade, and solidarity economies, which build the local economy, are based on paying producers full prices and recognizing the dignity of labor. That these just and honest economies are still small is due to the distortion of the global economy kept in place by coercive and undemocratic trade policies. These economies, while individually small, are significant when combined. The context must change to allow local economies to unfold and evolve and reach their full potential. Earth Democracy is based on diversity. It is based on multidimensional and multifunctional expressions of creativity and productivity in humans and nature. For local economies to grow they need governments performing their duties—as the governments of the Third World did in Cancun to stall the acceleration and expansion of corporate globalization. They need people mobilizing globally to protest, as they did in Seattle and Cancun. But most importantly, they need action at local and national levels to resist corporate globalization, and, through constructive action, build living economies. Living economies must grow like the seed to the tree, upward and outward. Genocidal and suicidal economies can be forced down on people and countries. Living economies are created with nature and with solidarity between people. They are built upon relationships of interdependence, unlike the dependency created by the mechanisms of corporate globalization, which are impoverishing the Third World and pushing farmers into debt and suicide.

Living economies are nonviolent economies; they are compassionate economies, unlike market economies based on violence and greed. Living economies can only grow with the nurturing and cultivation of living democracies.

Much debate has taken place about the alternatives to corporate globalization—and the debate is usually frozen in an either/or dichotomy. Being locked into fixed positions does not create movements. Building living democracy is not an either/or issue. Living democracies need to simultaneously devolve power to citizens and communities from both governments and institutions and make national governments, international institutions, and corporations more democratic and subject to social control. In fact, not only is building democracy at local, national, global levels not mutually exclusive, it is inseparable. Only a vigilant, vibrant citizenry can democratize national governments and global institutions. Living democracies reclaim from governments the rights and. responsibilities belonging to community and society, and reclaim from the global level what belongs at the national and local level. Only on the foundations of strong local democracy can strong national and global democracies be built.

There are several aspects which work to build living democracy:

Reinvent citizenship and reclaim the commons for the community. Reclaiming citizenship and the commons involves taking control over resources, livelihoods, and decision-making from the state as well as regaining resources that have been privatized by corporations, through the rules of the WTO and the structural adjustment programs of the World Bank and the IMF.

Reinvent government by enlarging people's sovereignty in order to protect livelihoods, resources, and the rights of citizens, and to regulate capital. Simultaneously define areas of intergovernment responsibility for regulating actions that impinge on the global ecological commons, such as the atmosphere. The dominant development model is based on centralized governments usurping the roles of local communities and citizens. We need to reinvent government to give back individuals and communities what be longs to society and take back from corporations and global institutions decision-making that is best carried out through national democratic processes. Centralized state control preventing farmers from saving seeds or centralized government control over issues of abortion are examples of government usurping the democratic and free spaces that should belong to people as individuals, families, and communities. Corporate globalization is dismantling the role of the state in regulating corporations, while religious fundamentalism is demanding that the state has an increased role in policing people. These invasions and perversions need to be corrected to make democracy real.

Reinvent global institutions and governance. Global institutions such as the World Bank, the IMF, and the WTO reflect neither the people's will nor their democratic expression. Global institutions are increasingly failing to defend the citizens of the world. They are becoming instruments of powerful corporations and nations. Meanwhile, global institutions are appropriating powers that appropriately lie at the local and national levels. The WTO has, for example, claimed rights to control biodiversity from local communities and national governments. This has imposed on local communities the privatization and enclosure of their seed, biodiversity, and knowledge commons. It has imposed on national governments laws that restrict the ability to shape patent laws appropriate to their socioeconomic context. Rules for management of biodiversity are most effectively shaped by local communities. Responsibility for laws on patents appropriately lie at the national level, so that citizens can influence these national policies to ensure that governments defend the rights of citizens to have access to seeds, food, and medicine. Redefining the role, function, and power of global institutions simultaneously rewrites the powers of global corporations, because corporate rule is established through global institutions exercising undemocratic power.

Because we live in an interconnected world, changes at the local level have global implications, and changes at the global level have implications for local economies, local cultures, and local democracy. Living. democracy connects the local, national, and global with a creative, coherent synergy that can bring about sustainability, justice, and peace. Since this synergy emerges from self-organization across our diversities, it grows without external control or domination.

The Community, the State, and the Corporation

Globalization has rendered the relationship between the community, the state, and the corporation totally fluid. The appeal of globalization is usually based on the idea that it involves less red tape, less centralization, and less bureaucratic control. Globalization is celebrated because it implies the erosion of the power of the state.

Most of the ideological projections of globalization have focused on the new relationship between the state and the corporation, the government and the market. Reflecting the ideology of deregulation, India's current prime minister and former finance minister Manmohan Singh stated, "Power should move to the boardroom."[4] The shift from the rule of the nation-state to that of the corporations does not give more power to the people. If anything, because corporations, especially transnational corporations, are more powerful and less accountable than governments, it puts less power in the hands of the people.

In recent years, the economic and political power of transnational corporations has been increasing dramatically. Transnational corporations control one-third of the world's wealth. Trade within firms, rather than between them, represents a growing proportion of world commerce.

Corporate globalization means less government regulation of business and commerce. But less government for commerce and corporations goes hand in hand with more government in the lives of people. As globalization increasingly transfers resources from public to corporate domain, discontent and dissent necessarily increase. Options for survival may only lie outside the law. As we saw in Bolivia, the theft of common resources through the privatization of water, for example, is condoned; collecting rainwater was forbidden. In such a situation, even a minimalist state restricted only to policing will become enormously large and all pervasive, devouring much of the wealth of society and invading every aspect of citizens' lives.

The erosion of the power of the nation-state concentrates power in the hands of corporations. It does not devolve power to the people; it does not move power downward into the hands of communities. It removes power from the local level and transforms institutions of the state from being protectors of the health and rights of people to protectors of the property and profits of corporations. This creates a state more committed to the protection of foreign investment than the protection of its citizens. The inversion of the state is well exemplified in a recently announced proposal that foreign security experts would train Indian police to protect the "life and property of foreign investors."[5] Another example is the Indian governments' recent decision to raise the permissible level of MSG (Mono Sodium Glutamate), which has been found to cause severe health problems such as asthma, in order to prioritize the expansion of fast food chains like KFC over people's health.[6]

Globalization's expansion runs into conflict with the democratic space needed for citizens to determine and influence the conditions for their health and well-being. The expansion of corporate control is often disguised as the expansion of democracy through consumer choice. However, choice within a predetermined set of options is not freedom. Corporate democracy requires the surrender of our right to determine the context of living and the values that govern society. The greater the scope of individual, elite consumer choice in automobiles and junk foods, the less ability communities have to control their local natural resources through a democratic public process.

From Politics of Exclusion to Politics of Inclusion

Corporate control or community control over natural resources: Which way points to a sustainable, just, and democratic future? The major issue emerging in India is the right to survival of the large number of poor people who derive their livelihood from com mon access to natural resources—land, water, and biodiversity. In each sector, a major conflict is emerging between corporate control and community control.

People's movements are demanding that power should not be concentrated in institutions of the centralized nation-states. Power should be distributed throughout society and should be dispersed through a multiplicity of institutions, with most power exercised at the local level by local communities. However, the transnational corporation-driven agenda requires that power move from the centralized control of nation-states to the even more centralized control of global corporations and global institutions like the WTO and the World Bank and the IMF.

People are redefining democracy in terms of decisions in their everyday lives. They are redefining the nation in terms of people, not in terms of a centralized state. This trend toward localization was, in fact, born alongside globalization. If globalization is the corporate-driven agenda for corporate control, localization is the countervailing people's agenda for protecting the environment, survival, and livelihood. In the absence of regulation by national governments, people are creating their own ecological and democratic response.

Living democracy is based on the living diversity of cultures and communities but also on the idea that we all share one common humanity and one commonality with all beings and life forms. It is simultaneously local and global and therefore transcends the exclusionist logic of either/or, and evolves through the non-duality and non-separability of relationships. Relationships create the space for people to respond to one another. This creates responsibility and establishes the grounds for sharing and compassion. Relationships exist because *we* all belong to an earth family—through sharing the same biosphere and living on the same earth, one planetary home, we are all relations. This is our most fundamental identity. It is shaped both by the particular place on earth where we live our everyday lives, as well as the planetary consciousness of being members of one earth family, one humanity.

Localization is Not Autarky

Localization does not imply autarky or insularity. It involves subjecting the logic of globalization to the test of sustainability, democracy, and justice in each concrete instance of foreign or large-scale investment. It also involves reclaiming the state as a tool for the protection of people's interests and returning to ordinary people the power and authority the state has usurped. In living democracy, the core of government is self-government and self-rule. Only areas of life that do not lend themselves to self-organization need government. The real issue of our times is how to reinvent the state in a way that is not centralized, bureaucratic, and controlling, in a way grounded in the community and responsible to community. Leaving decisions on the distribution of goods and services and on environmental impact to unregulated and nonaccountable market forces would also be an error. Social regulation of the market requires strong community rights and social policies—and this is not the same as individual consumer choice. The contest between the transnational corporations, the force behind globalization, and the citizens and local communities, the force behind localization, spins off into a contest over what kind of state will regulate corporations while recognizing and enhancing freedom for people. Freedom from want—from hunger and homelessness and the denial of basic needs—is the most fundamental freedom, without which there can be no other freedoms. Ensuring this freedom by building living democracies, strengthening civil society, and empowering people is the project of democracy in our times.

People's Protectionism

Movements for localization are giving rise to a new people's protectionism. The power and authority to make environmental and economic decisions are moving from centralized states to self governing structures at the local level. Citizens and community organizations are now deciding which roles and functions the state should have. People are seeking to transform the institutions of society—the courts, police, government departments—currently distorted to protect the interests of transnational corporations and sacrifice the interests of citizens, small producers, and small traders.

The combination of strong-arm tactics of the Super 301 clause of the US Trade Act, the IMF and the World Bank pressures for liberalization, and the Uruguay Round of GATT have been used by the largest corporations of the world to force new investment opportunities in India since 1991. These investments and policies have combined to put the interests and rights of foreign investors above the interests and rights of the citizens of India. But now, in each sector, the biggest multinational corporations have been forced to recognize that clearance from citizens, not just the government, is necessary for democratic functioning.

Whether it is Coca-Cola in Kerala; Suez in Delhi; Monsanto, Cargill, and W. R. Grace in Karnataka; DuPont in Goa; or KFC in Delhi and Bangalore, the entry of transnational corporations is being questioned by local communities and grassroots struggles. Local communities are raising a common voice to say, "We will decide the pattern of investment and development. We will determine the ownership and use of our natural resources." As this message resonates in village after village, from one investment site to another, a new environmental philosophy based on democratic decentralization and political and economic localization is put into practice. The pressure of the people is forcing the government to remember its role as protector of the public interest and the country's natural and cultural heritage, not merely the interests of foreign investors.

A pattern is emerging for environmental governance beyond the centralized state and superstate systems, which work for the corporate interests. Localization is emerging as an antidote to globalization and to unrestrained commercial greed.

Diversity and Freedom

The Green Revolution is an exemplar of the deliberate destruction of diversity. New biotechnologies are now repeating and deepening these tendencies. Furthermore, the new technologies, in combination with patent monopolies being pushed through intellectual property rights regimes in GATT, the biodiversity convention, and other trade platforms, are threatening to transform the diversity of life forms into mere raw material for industrial production and profits. They simultaneously threaten the regenerative freedom of diverse species and the free and sustainable economy of small peasants and producers based on nature's diversity.

The seed, for example, reproduces itself and multiples. Farmers use seed both as grain and for the next year's crop. Seed is free, both in the ecological sense of reproducing itself, and in the economic sense of reproducing the farmers' livelihood.

This seed freedom is a major obstacle for seed corporations. To create a market for seed the seed has to be transformed materially so that its reproductive ability is blocked. Its legal status must also be changed so that instead of being the common property of farming communities, it becomes the patented private property of the seed corporations.

The seed is starting to take shape as the site and symbol of freedom in the age of manipulation and monopoly of life. The seed is not big and powerful, but can become alive as a sign of resistance and creativity in the smallest of huts or gardens and the poorest of families. In smallness lies power. The seed also embodies diversity. It embodies the freedom to stay alive. Seed freedom goes far beyond the farmer's freedom from corporations. It represents the freedom of diverse cultures from centralized control. In the seed, ecological issues combine with social justice. The seed could play the role of Gandhi's spinning wheel—a symbol of freedom during India's independence movement—in this period of recolonization through free trade.

In cooperation with the movements I have been working with over many years and overcoming blocks and pressure, I launched a national program to save seed diversity in farmers' fields. We call it Navdanya, which literally means nine seeds and is a beautiful symbol of the richness of diversity. Ours was not the first seed conservation program. Genetic resources have always been collected for breeding. The risks of breeding toward uniformity led to the emergence of government gene banks in the 1970s. However, while gene banks collect biodiversity from farmers' fields, they do not conserve it through and with farmers. Instead, diversity flows from farmers' fields to gene banks and then on to corporate breeders. The farmers become mere consumers of corporate seed. This excludes the farmer from the critical role of conserver of genetic diversity and innovator in the utilization and development of seed. It robs farmers of their rights to their biological and intellectual heritage. It separates consecration from production and scientists from farmers. Navdanya wanted to build a program in which farmers and scientists relate horizontally rather than vertically, in which conservation of biodiversity and production of food go hand in hand, and in which farmers' knowledge is strengthened, not robbed.

While the fundamental changes we are working toward can only be achieved in the long term, Navdanya has already had major impacts in the villages in which we work. Realizing that our small efforts to conserve indigenous seed diversity are not enough, we have also joined hands with the farmers' movement to urgently mobilize public opinion against the emerging threat of multinational corporations gaining monopoly control on all life through new biotechnologies and intellectual property rights.

In 1991, I started to contact the farmers' organizations, to alert them about new trends, to work with them on protecting farmers' rights to freely conserve use, exchange, and modify the seeds. In February 1992, we organized a national conference on GATT and agriculture with the Karnataka Rajya Ryota Sangha (KRRS). In October 1992, at a massive farmers' rally in Hospet organized by the KRRS, the seed satyagraha was launched, following Gandhi's politics of satyagraha as a fight for truth based on non-cooperation with unjust regimes. In March 1993, we held a nationally rally in Delhi at the historic Red Fort under the leadership of the national farmers' organizations, the Bharatiya Kisan Union. Independence Day, August 15, 1993, was celebrated with farmers asserting their *samuhik gyan sanad* (collective intellectual property rights). On

October 2, 1993, one year of the seed satyagraha was celebrated in Bangalore with a gathering of 500,000 farmers. We also had farmers from other Third World countries, as well as scientists who work on farmers' rights and sustainable agriculture in an expression of solidarity. The internationalization of the seed satyagraha had, within one year, given the word *globalization* a new meaning. No longer did globalization represent global markets, as in the parlance of free trade proponents, it had come to mean the globalization of people's resistance to centralized control over all aspects of their life. In 1994 we began the neem campaign against the patenting of neem by W. R. Grace. We mobilized to prevent the government from signing the GATT agreement in Marrakesh in 1994, and when the WTO still came into being, we organized to continue to question and challenge it. Internationally we organized as the International Forum on Globalization in 1994 and, after Seattle, as the "Our World is Not for Sale" Network. In India, 250 organizations came together in the People's Campaign Against the WTO. The campaign was convened by S. P. Shukla, the former ambassador to GATT, who had negotiated the Uruguay Round and knew the tricks that had been used to impose undemocratic "agreements" on people and included four ex-prime ministers, trade unions, women's movements, farmers' unions, and environmental movements. In 2000, the movements fighting for people's natural rights to natural resources gathered where Mangal Pandey launched the first movement for India's independence in 1857. We committed ourselves to defend and reclaim our fundamental freedoms related to land, forests, biodiversity, food, and water. That is how the movements for *bija swaraj* (biodiversity and seed democracy), *anna swaraj* (food democracy), and *jal swaraj* (water democracy) were born.[7]

Native seed has become a method of resisting monocultures and monopoly rights. The sustainable shift from uniformity to diversity respects the rights of all species. Protecting native seeds is more than simply conserving raw material for the biotechnology industry. The seeds being pushed to extinction carry within them seeds of other ways of thinking about nature, and other ways of fulfilling our needs.

Conservation of diversity is, above all, the commitment to let alternatives flourish in society and nature, in economic systems, and in knowledge systems. Cultivating and conserving diversity is no luxury in our times. It is a survival imperative. It is the precondition for freedom for all. In diversity, the smallest has a place and role, and allowing the small to flourish becomes the real test of freedom.

Seed Saving: Our Ethical Duty, Our Human Right

Seed is the first link in the food chain. In Sanskrit, *bija*, the seed, means the source of life. Saving seed is our duty; sharing seed is our culture.

Patents on seeds and genetic resources rob us of our birthright and deprive us our livelihoods by transforming seed saving and seed sharing into intellectual property crimes. This is an assault on our culture, on our human rights, and on our very survival.

Poor peasants of the South cannot survive seed monopolies. That is why the case of one Canadian farmer, Percy Schmeiser, will decide not only his own fate, but that of billions of peasants. Schmeiser's canola crop was contaminated by Monsanto's Round Up ready canola. Instead of compensating Schmeiser for polluting his crop, Monsanto sued for "intellectual property theft."[8] The unjust and unethical case brought by Monsanto against Schmeiser is a double crime against farmers. First, by creating and enforcing illegitimate patent rights to seed, it robs us of our human right and human duty to be seed savers. Second, it rewards the polluter with enhanced property rights and profits. The principle of "polluter pays" has been transformed into "polluter gets paid."

This perverse jurisprudence must be corrected for the sake of all farmers, and all species. Farmers' freedoms must come before those of corporate monopolies. Farmers' survival must come before corporate greed. Schmeiser's future is our future. His seed freedom is our freedom. His rights as a farmer are symbolic of the human rights of all farmers.

Gandhi's creative vision of swadeshi, swaraj, satyagraha and *sarvodaya* (inclusion) inspires us to build living economies and living democracies. In his legacy we find hope, we find freedom, we find our own creativity. Gandhi's philosophy charges our actions with life. Through Gandhi we begin with constructive action and turn it into our best resistance. When governments begin to implement TRIPS legislation, as the Indian government has done through three amendments to the Patent Act and by creating new plant-variety legislation, we remember Gandhi's words, "As long as the superstition that people should obey unjust laws exists, so long will slavery exist," and we renew our commitment to the bija satyagraha. Mahatma Gandhi started the salt satyagraha to protest the Salt Laws and the colonization of salt by the British Empire. In this tradition, the people's movement in India committed to the bija satyagraha to resist and pledge noncooperation with unjust and immoral intellectual property laws.

As genetic pollution threatens our biodiversity and globalization threatens our farmers, we create living economies and living democracies based on swadeshi and swaraj. Just as the seed has the potential to germinate and evolve and renew itself perennially, Gandhi's legacy has the potential to germinate, evolve, and renew our actions and strategies for freedom appropriate to our times and context.

From the seed, our swadeshi efforts have grown to incorporate *jaiv kheti* (organic farming) and fair and just trade. Swadeshi, in the form of biodiversity conservation, has evolved organically into the swaraj of *jaiv panchayat* (living democracy), resting on the resistance of satyagraha—noncooperation with immoral, unjust laws.

Gandhi's legacy lives and gives us hope to shape ever new instruments to keep life in its diversity free. Gandhi's legacy carries the seeds for the freedoms of humans and all species. Gandhi's legacy is humanity's hope.

Living Democracy Movement

The success of movements at Seattle and Cancun, and the amazing mobilizations for the World Social Forums are examples of an emergent politics based on diversity and self-organization, not on monocultures and manipulation. They have thrown up a new model of self-organizing as democracy. The success on both the local and global levels is indicative of the potential for self-organization as a basis for transforming politics at all levels.

The violence of corporate globalization on the one hand, and wars justified on grounds of shallow religions and narrow nationalist identities on the other, demand a response that is simultaneously local and universal. Local responses include reducing our ecological footprint and creating livelihood and job opportunities. Our diverse identities, rooted in biological and cultural diversity, our sense of place, and our sense of belonging also emerge from our local response. Universal responses spring from the acknowledgement that we share life with the rest of life, and our humanity with all of humanity.

The dominant form of corporate globalization takes a narrow, highly localized interest, and imposes it as the universal. This imposition involves deep structural violence and triggers vicious cycles of violence as identities are threatened, securities are eroded, and a backlash emerges in the form of "terrorism." The universal cannot be a globally imposed local interest. It is the emergent quality of all people living by the universal principles of nonviolence—the ecological sustainability of nonviolence to non-human life and the social and economic justice of nonviolence to human life. The universal is the unfolding of the potential of diverse and multiple locals, acting in self-organized ways but guided by the common principles of love and reverence for life. Tolstoy wrote from his deathbed:

> Understanding that welfare for human beings lies only in their unity, and that unity cannot be attained by violence. Unity can only be reached when each person, not thinking about unity, thinks only about fulfilling the laws of life. Only this supreme law of love, alike for all humans, unifies humanity.[9]

One of the results of the Cartesian, mechanistic worldview allows for the violent imposition of one's position on others, with the conviction that this is for the good of the other. The Iraq war is supposed to have been good for the Iraqis. On the other hand, this view of the mechanistic universal makes ordinary people shrink from taking initiative for change, because in a mechanically de fined unity, it is either all or nothing.

As Gandhi observed:

> It is necessary for us to emphasize the fact that no one need wait for anyone else in order to adopt a right course. Men generally hesitate to make a beginning, if they feel that the objective can not be had in its entirety. Such an attitude of mind is in reality a bar to progress.[10]

The living democracy movement is based on the acknowledgement that we can begin where we are and that we can imbue our everyday actions with the broadest of visions, the deepest of values. The principles of Earth Democracy—that we are members of the earth family, our deepest identity is our earth identity, and our highest duty is to protect all life on earth—grew from the jaiv panchayat movement.

On August 9, 1999, hundreds of village communities organized as jaiv panchayats served notice to the director general of the WTO, Mike Moore, as part of their campaign against biopiracy. The letter read, in part:

> We wish to inform you that we will not allow you to take decisions on matters that fall exclusively within our jurisdiction through our decentralized democratic system. On the basis of our inalienable rights that are

> recognized by our Constitution· and the CBD [Convention on Biological Diversity—a UN Agreement signed at the Earth Summit], we will not permit WTO to undermine our rights and protect those who steal our knowledge and our biodiversity.

Mike Moore came to India in response. And the government of India had to acknowledge the problem of imposing TRIPS on local communities in its submission to the WTO. More recently, local mobilizations have responded to the undemocratic imposition of genetically modified organisms (GMOs) through the WTO. The GMO challenge is a global response of local and national citizens movements to the US-initiated WTO dispute against Europe.

The living democracy movement is based on a local-global, micro-macro symbiosis. Navdanya could not have been started without the knowledge or inspiration provided by the 1987 Laws of Life Conference. And the movements to resist TRIPS and GMOs at the global level would not have emerged in the form they did without the new possibilities and potentials which opened up ecological farming methods as superior alternatives to genetic engineering, golden rice, protein potatoes. Our work to articulate and defend biodiversity and knowledge as a living commons through common intellectual rights and community seed banks has created alternatives, not just for local communities but for all societies. We refused to allow an enclosure of our last freedoms. And our resistance has opened up spaces in different spheres and different places for others.

From Biodiversity to Monocultures

Industrial agriculture based on high external inputs of chemicals and water creates a push for uniformity and monoculture and leads to the erosion of biodiversity. Each agroclimatic zone has evolved farming systems based on species adapted to it. Industrial agriculture destroys ecosystem and farming diversity. It also pushes crops to extinction. Thus rice and wheat monocultures have replaced diverse millets, pulses, and oilseeds, often grown as mixtures and in rotation. Finally, industrial farming destroys diverse varieties of crops and replaces them with uniform varieties adapted to chemicals, not to ecosystems and climate.

Over centuries, Indian farmers selected and cultivated thousands of varieties of rice from a wild aquatic grass. No other cultivated crop has been developed to such an extent. There are varieties of rice to fit thousands of ecological niches all over the country, from the temperate high hills of the Himalayas to the tropical lowlands to deep-water and salt-water marshes of the sea coasts. The varietal diversity of cultivated rice in India can be considered to be the richest in the world, with the total number of varieties estimated to be around 200,000.

India's rices possess a wide diversity of morphological and physiological characteristics. They range in duration from 60–200 days to attain maturity and are adapted to several different ecological conditions. There are varieties which are completely purple in color. The grain length varies from 3.5 to 14 millimeters and in breadth from 1.9 to 3 millimeters. The grain color and quality also vary from red rice to fine and elongated white rice, from scented to nonscented rice and glutinous rice. The extent of diversity is evident from the fact that over 1,500 morphologically distinct varieties have been found cultivated in the Jeypore tract of Orissa. Each local indigenous rice variety, over the years, has adjusted to the ecosystem of that particular region, including the environmental fluctuations and vagaries of monsoon. Thus, even under unfavorable conditions, a minimum yield is assured. These traditional varieties have been selected by farmers over thousands of years for their desired characters, such as taste, cooking qualities, aroma, and medicinal properties.

In their classic *Wheats of India*, Sir Albert Howard and his wife G. C. C. Howard, with Habibue Rehman Khan, identified thousands of wheat varieties belonging to 10 subspecies. Today, in most regions, only a handful of varieties bred to respond to chemicals are being grown.

Is this loss of diversity necessarily a bad thing? Modern agriculture views this genetic uniformity as advantageous for securing greater yields and for developing specifically desired morpho-agronomic cultivators. Typically, in developing or adapting a variety in response to the pressure to get something on the market quickly, scientists will search for one major gene to confer resistance on. However traditional resistance is not so simple. Resistance may be the product of many genes working together. However, breeding in this kind of complex resistance is too time consuming, complicated, and costly to the modern breeder.

By utilizing one-gene resistance, the plant breeder gives pests or disease an easy target, with only one line of defense to get around. This danger of uniformity has been witnessed quite a few times in history: the Irish potato famine of the 1840s; the pearl millet wipeout by downy mildew, which led to starvation in India in 1971; and the recent crisis in the rice fields of South Asia due to the widespread rice stunt virus infection.

The virtue of possessing traditional varieties was demonstrated in Zambia, where 90 percent of the domestic supply of maize was grown through the use of a highly uniform hybrid maize cover. In 1974, however, a new mold attacked the crop, and 20 percent of the hybrids were infested while the impact on traditional maize was negligible.[11] This is a concrete example of the strength of natural diversity. Recent studies in West Bengal show a dramatic increase in rice pests and diseases after the introduction of high yielding varieties.

Thus, for many farming communities, diversity—be it social, cultural, or genetic—means security. Genetic diversity provides security for the farmer against pests, disease, and unexpected climatic conditions. It also helps small-scale farmers to maximize production in the traditionally highly variable environments in which they cultivate their crops. Higher yields are obtained from employing a mixture of crops and crop varieties, each one specifically adapted to the microenvironment in which it grows, rather than by using one or a few "modern" varieties. Such uniform varieties will only reach their potential if the environment is also uniform. That means high-quality land, where fertility and water status have been evened out with the use of fertilizers and irrigation, conditions and inputs generally unavailable to the small-scale farmer. Even when available, the use of inorganic fertilizers has led to the growing crisis of soil erosion and depleted soil quality. Likewise, year-round irrigation has led to a severe depression of the water table. Major irrigation projects wrought by big dams on rivers have caused more ecological and social problems than they have solved.

Additionally, genetic diversity provides farming communities with a range of products with multiple uses. Some varieties of a particular crop may be good for immediate consumption, others for long-term storage, yet others for pest resistance, and so on. This genetic wealth is, in addition, an important reservoir of diversity for agriculture worldwide, providing crucial characteristics for pest and disease resistance, nutritional quality, and other factors.

Based on thousands of years of experience and a deep knowledge of their needs and their agricultural production systems, communities have developed multiple strategies for their farming systems, like intercropping and agroforestry. Traditional farming systems use at least several varieties of each crop. For cereals like sorghum, rice, wheat, and barley, which are self-pollinated, and for vegetatively propagated crops like potatoes and bananas, the number of varieties used may be very high.

Globally there is a growing ecological awareness among consumers and a greater demand for chemical-free food. Local farmers' varieties, which have evolved free of chemicals, are the best choice for organic farming and other alternative agricultural practices.

Toxic Pollution

Industrial farming is based on the use of toxic chemicals such as pesticides and herbicides. These chemicals were designed, like Agent Orange, as weapons for mass destruction. The Bhopal disaster is a tragic reminder of the hazards of toxic agrochemicals. These chemicals are produced because the new varieties are prone to pests and disease.

Whenever the new varieties of rice have been introduced in Asia by the International Rice Research Institute, the global gene bank for rice, they have proven to be susceptible to diseases and pests. IR-8 was attacked by bacterial blight in Southeast Asia in 1968 and 1969.[12] In 1970 and 1971, it was destroyed by the tungro virus. In 1975, half a million acres under the new rice varieties in Indonesia were destroyed by pests. In 1977, IR-36 was developed to be resistant to eight major known diseases and pests including bacterial blight and tungro. But this was attacked by two new viruses, ragged stunt and wilted stunt.

In Punjab, the experience with the new varieties was no better. They created new varieties of pests and diseases. The Taichung Native I variety, which was the first dwarf variety introduced in 1966, was susceptible to bacterial blight and white-backed plant hopper. In 1968 it was replaced by IR-8, which was considered to be resistant to stem rot and brown spot, but proved to be susceptible to both. Later varieties such as IR-103, IR-106, IR-108, and IR-109, which were released after the failure of earlier dwarf varieties, were specially bred for disease and insect resistance. IR-106, which currently accounts for 80 percent of the rice cultivation in Punjab, was considered resistant to white-backed plant hopper and stem rot disease when it was introduced in 1976. It has since become susceptible to both, as well as to the rice leaf folder, hispa, stem borer, and several other insect pests. The miracle varieties have eroded the diversity of traditionally grown crops and have become a mechanism for introducing and fostering pests.

According to the Indian Council for Agricultural Research, only 1 percent of pesticides actually reach the target pest with the remainder impacting non-target sectors. It has also been estimated that despite heavy pesticide use, pests are now causing damage to some 35 percent of the crops, compared with the pre-pesticide era rate of 5 to 10 percent. The number of damaging insects is also increasing. The estimated number of insects damaging rice paddies increased from 40 in 1920 to 299 in 1992. Likewise, it increased for pulses from 10 to 240 and for wheat from 10 to 120.[13] Nevertheless, the development and use of pesticides continues to increase.

The recent report of the joint parliamentary committee on pesticides in soft drinks shows how pervasive these toxins have become by reporting on the high percentage of pesticides found in all varieties of Coca-Cola and Pepsi beverages. An environmental problem has been transformed into a public health emergency.

Pollution and Depletion of Water Resources

Industrial agriculture is water-wasting, water-depleting, and water polluting. Crops bred for chemical farming need 5 to 10 times more water than crops bred for ecological farming. In addition, the preference given to water-intensive crops such as intensively irrigated paddy, wheat, and sugarcane has led to the displacement of water-conserving, high-value, high-nutrition crops such as rain-fed wheats (kathia, mandua) and water-prudent crops like millets. Chemical farming increases water demand while simultaneously reducing the water-conservation capacity of soils by lessening the return of organic matter to the soil. Intensive irrigation has also led to severe problems such as waterlogging (the creation of a wet desert) and salinization (the rising of salts to the surface).

Erosion of Soil and Soil Fertility

Industrial agriculture is leading to severe erosion of soil and soil fertility. For centuries, soil fertility was maintained in India by good farming practices. The rich alluvial soils of the Indo-Gangetic plains of North India have been farmed for centuries without depleting soil fertility. Of those soils, eminent agricultural scientist Howard and Wad said:

> Field records of ten centuries prove that the land produces fair crops year after year without falling in fertility. A perfect balance has been reached between the manorial requirements of the crops harvested and the natural processes which recuperate fertility.[14]

And in his presidential address to the agriculture section of the Indian Science Congress, G. Clarke said:

> When we examine the facts, we must put the Northern Indian cultivator down as the most economical farmer in the world as far as the utilization of the potent element of fertility, nitrogen, goes. He does more with a little nitrogen than any farmer I ever heard of. We need not concern ourselves with soil deterioration in these provinces. The present standard of fertility can be maintained indefinitely. [15]

However, centuries of conservation have been undone by five decades of careless farming. Monocultures have increased exposure of soils to wind and rain, aggravating erosion. Chemical fertilizers have undermined the soil fauna and flora that create and maintain soil fertility.

Greenhouse Gases and Climate Change

Industrial agriculture, with its chemical-intensive and fossil-fuel intensive inputs, is responsible for large contributions of green house gases. It is responsible for 25 percent of the world's CO_2 emissions, 60 percent of its methane gas emissions, and 80 percent of its nitrous oxide emissions, all powerful greenhouse gases.

Nitrous oxide is 200 times more potent than CO_2 as a greenhouse gas. It is produced by the use of nitrogenous fertilizers. Around 70 million tons of nitrogen fertilizers are used in agriculture each year, producing 22 million tons of nitrous-oxide emissions.

Emissions of carbon from the burning of fossil fuels for agricultural purposes in England and Germany are as much as 0.046 and 0.053 tons per hectare. Such emissions are roughly seven times lower in nonindustrial agriculture so the industrialization of agriculture will lead to increased carbon emissions worldwide.[16]

Atmospheric pollution due to greenhouse emissions is aggravated by drought and floods. This climate instability is a threat to agriculture and food security. A shift to ecological agriculture and organic farming is an ecological, economic, and security imperative.

Organic Farming: The Ecological and Economic Imperative

Industrial agriculture has been promoted, financed, and subsidized in spite of the high cost to the environment. The argument used is that these ecological costs are a necessary part of increasing productivity. However, the productivity of industrial agriculture is actually negative. More resources are used as inputs than are produced as outputs. Usually productivity is increased by the implementation of labor displacing machinery and chemicals. However, labor is not the scarce input. Land and water are. If, instead of focusing on labor costs, we take energy, natural resources, and external inputs into account, then industrial agriculture does not have higher productivity than ecological alternatives. Over the last 50 years, the shift from internal input to high external input agriculture has resulted in a sixty six fold decrease in productivity.[17] We need to make an ecological transition to produce more food using fewer resources.

This productivity analysis is based on a study comparing traditional polycultures with industrial monocultures and shows that a polyculture system can produce 100 units of food from 5 units of inputs, whereas an industrial system requires 300 units of input to produce the same 100 units. The 295 units of wasted inputs could have provided 5,900 units of food. This is a recipe for starving people, not for feeding them.

A common argument used to promote industrial agriculture is that only it and industrial breeding can maintain the increased food productivity needed for a growing population. However, since resources, not labor, are the limiting actor in food production, it is resource productivity, not labor productivity, which is the relevant measure. What is needed is more efficient resource use so that the same resources can feed more people. A 66-fold decrease of food producing capacity in the context of resources use is not an efficient strategy for using limited land, water, and biodiversity to feed the world.

Not only is the measure of productivity of industrial agriculture incomplete because all inputs, including resource and energy inputs, are not taken into account, but also because not all outputs are taken into account.

Ecological agriculture is based on mixed and rotational cropping, and the production of a diversity of crops.

Traditional agricultural systems have evolved polycultures because a greater yield can be harvested from an area planted with diverse crops than from an equivalent area consisting of separate patches of monocultures. For example, by planting sorghum and pigeon pea mixtures, one hectare will produce the same yield as 0.94 hectares of sorghum monoculture and 0.68 hectares of pigeon pea monoculture. Thus, one hectare of polyculture produces what 1.62 hectares of monoculture can produce. This is called the land equivalent ratio.

Small biodiverse farms increase both output and incomes of small farmers. Small farms in West Bengal growing 55 different crops gave incomes of 227,312 rupees per acre; a farm with 14 crops gave 94,596 rupees while a monoculture farm brought in only 32,098 rupees per acre.

Contrary to the dominant myth that monocultures and industrial farming is necessary for producing more food, and hence, biodiversity must be destroyed to solve the problem of hunger, value and returns from farm produce increases significantly with greater diversity of crops.[18]

From Dying Democracies to Living Democracies

Negative economies and negative politics feed on and fuel negative cultures and identities. Cultures have been shaped by the land and cultural diversity has coevolved with biological diversity. Cultures have shaped positive identities—based on a sense of place in eco systems and economies. As people are displaced and insecurities grow, identity is transformed and destroyed. Among these negative cultures and identities, terrorism, extremism, and xenophobia take virulent form. Humanity defines itself through its inhumanity. Vicious cycles of violence and exclusion—cultural, political, economic—predominate.

Our survival demands that we make a transition from vicious cycles of violence to virtuous cycles of nonviolence; from negative economies of death and destruction to living economies that sustain life on earth and our lives; from negative politics of corruption and fascism to living democracies which include concern for and participation of all life; and from negative cultures that are leading to mutual annihilation to positive and living cultures based on caring, compassion, and conservation.

Earth Democracy allows the emergence of living economies, living democracy, and living cultures.

Globalization is threatening survival itself—by robbing millions of their right to life and by creating a political climate in which negative identities thrive. Human rights must focus on the right of the human species to survive, in peace with each other and the rest of the earth family. Economic globalization does not create global markets, it creates global madness. This madness must be stopped. With our collective will and our courageous interventions we must cure not the symptoms of insecurity but the root causes.

It is essential to dispel the illusion that globalization is natural and inevitable. Globalization is a political project and it needs a political response. Our political response needs to put human beings, in all our diversity, at the center of economic thought. We must not allow the annihilation of human rights by all-powerful global corporations. We must stop treating corporations, markets, and capital as people for whose protection all human beings can be put at risk.

And we need to evolve an agenda for human rights which includes all humans and all rights. Most liberation movements in recent history have been partial and exclusivist. They worked for a class, a race. And often they were based on violence. They excluded other species, they excluded diverse cultures, and they frequently excluded women's politics of making change through everyday life. We have an opportunity today to seek freedom in inclusive ways, in our diversity, to seek freedom for humans in partnership with other species, and to seek freedom nonviolently. This freedom of diversity is the alternative to globalization.

Globalization has pushed representative democracy to its final test. The combination of corporate globalization and electoral democracy is separating leaders and governments from society and people. Whether governments are pushing the globalization agenda despite popular resistance, the militaristic agenda against the democratic will of people, or the divide-and-rule agenda of religious fundamentalism and xenophobia, they are no longer governing for economic justice, peace, or social harmony. While the democratic divide between people and their leaders is obvious, the next steps. for reclaiming democracy are being shaped by people, not by so-called leaders.

Living democracy for us has become a process for building alternatives while taking back power.

CHAPTER THREE

Living Cultures

Our age is marred by multiple forms of violence—economic, military, and cultural. The violence of the dominant culture is reinforced by the violent responses of those whose lands and cultures are invaded. A nonsustainable economic system based on principles of free trade, greed, and imperialism is creating vicious cycles of violence from which there seems to be no way out.

A few years ago, during a public dialog entitled "Globalization and Violence" in Udipi, South India, in reaction to the phrase "culture of violence," Samdhong Rinpoche, the prime minister of the Tibetan government in exile, said violence can not have a culture. The word culture in Sanskrit—*sanskriti*—means activities that hold a society and community together. Violence breaks societies up, it disintegrates instead of integrates. The practice of violence, therefore, cannot be referred to as "culture." In Sanskrit and Hindi, in fact, we have a word for destructive processes—*vkriti,* that which disintegrates and violates. In English, however, the phrase "culture of violence" is frequently used. During British rule in India, Gandhi was asked what he thought of Western civilization. He responded "It would be a good idea." An imperialist West cannot be a civilized West, since civilized people do not destroy other civilizations and cultures. To be civilized is to live and let live, both at the individual and societal level.

Imperialism has always operated under the pretense of making other cultures civilized while, in fact, destroying other cultures and robbing people of their humanity, diversity, and identity. Living cultures are based on cultural diversity and recognize our universal and common humanity. Killing cultures are based on imperialistic universalism—a violent imposition of the cultural priorities of an imperial power. The universal order of globalization and imperialism is not based on universal responsibility, compassion, and solidarity, but on the conquest and colonization of resources, of history, and of the past and the future.

False universalisms lead to war and violence; true universalisms based on our common humanity, our oneness, and our inter connectedness provide the conditions for peace, cooperation, and coexistence. Diversity and autonomy are treated as a problem and disease in the false universalism of imperialism, corporate globalization, crusades, and jihads, but in the universalism that creates peace, they are expressions of freedom.

In the eyes of imperialists, as non-Western cultures are invaded and conquered, their diversities and traditions disappear and the world is reshaped in the image of the colonizer—and the colonized will feel grateful for their "liberation." This used to be called the white man's burden. It continues in the idea of bringing "democracy" and a war named "Operation Iraqi Freedom."

The imperialists do not recognize their inability to respect the autonomy, self-organization, agency, and integrity of the other, creates diverse forms of violence.

Corporate globalization has unleashed a war against farmers, against women, against other species, and against other cultures. While the project of corporate globalization is based on the imposition of a global monoculture—a food monoculture shaped by McDonald's, Monsanto, and Coke; a dress monoculture; a media monoculture; a transport monoculture—we are not witnessing a disappearance of diversity. Diversity is becoming dominant. There are, however, two kinds of movements for cultural diversity that are growing.

One is the extremist and exclusivist "Talibanization" of culture—a patriarchal, militarist response to the empire that mimics the violence of the empire. While resisting imperial occupation, it simultaneously declares cultural wars within its boundaries—against women, minorities, and other groups. In Taliban-ruled Afghanistan, the violence against women and the destruction of the Bamiyan Buddhas are examples of destruction of culture in the name of the protection of culture. The reduction of the 2004 US elections to "red" and "blue" belts and conflicts over so-called cultural values are other examples of the destruction of diversity and pluralism through the construction of exclusivist identity. Such identifications give rise to the culture wars, crusades, and jihads of our times.

The other movement for cultural diversity can be found in the movement for peace, sustainability, and justice, which protects diversity through care and compassion, not through domination and conquest. These positive, diverse identities provide alternatives to the imperialistic, patriarchal models of relating to the diverse other. His Holiness the Dalai Lama has articulated how compassion and respect for human rights of people of all cultures can be a basis for public life and international relations:

> To me it is clear that a genuine sense of responsibility can result only if we develop compassion. Only a spontaneous feeling of empathy for others can really motivate us to act on their behalf.
> Democracy is the system which is closest to humanity's essential nature. Hence those of us who enjoy it must continue to fight for all people's right to do so.... We must respect the right of all peoples and nations to maintain their own distinctive characters and values.[1]

This philosophy of diversity plus universal responsibility provides the basis for cultivating living cultures from the midst of killing cultures.

From Cultures of Death to Cultures of Life

Why are we as a species destroying the very basis of our survival and existence? Why has insecurity been the result of every attempt to build security? How can we as members of the earth community reinvent security to ensure the survival of all species and the survival and future of diverse cultures? How do we make a shift from life-annihilating tendencies to life-preserving processes? How do we, from the ruins of the dominant culture of death and destruction, build cultures that sustain and celebrate life?

When reality is replaced by abstract constructions created by the dominant powers in society, manipulation of nature and society for profits and power becomes easy. The welfare of real people and real societies are replaced with the welfare of corporations. The real production of the economies of nature and society is replaced by the abstract construction of capital. The real, the concrete, the life-giving is substituted for by artificially constructed currencies.

Closely linked to the rule and reification of abstraction are the monoculture of the mind and the law of the excluded middle, which threaten life in its diversity, self-organization, and self-renewal.

The monoculture of the mind is the reductionist perspective which sees and constructs the world in terms of monocultures. It is a mind, blind to diversity and its richness, that pushes to oblivion and extinction biological and cultural diversity—the very preconditions of ecological and cultural security.

The law of the excluded middle, which is based on an either/or logic, becomes the basis of legitimizing exclusion and ecocide and genocide. It constructs the world in mutually exclusive categories, thus banishing multiplicity and pluralism as well as relationships and connectedness. It shuts out spaces between nature and culture. It denies the existence of biodiversity on farms and food from forests. It denies cultural diversity in our knowledge, our food, and our dress.

Even while the market economy erodes nature's economy and creates new forms of poverty and dispossession, the market is proposed as a solution to the problem of ecologically induced poverty. Such a situation arises because the expansion of the market is mechanically assumed to lead to development and poverty alleviation. In the ideology of the market, people are defined as poor if they do not participate overwhelmingly in the market economy and do not consume commodities produced for and distributed through the market. People who satisfy their needs through self-provisioning mechanisms are perceived as poor and backward.

Cultural perceptions which prejudice the market economy also impact this situation. As Rudolf Bahro observed, culturally conceived poverty based on non-Western modes of consumption are often mistaken for misery and poverty.[2] People are perceived to be poor if they eat millet or maize, common non-Western staple foods that are nutritionally far superior to processed foods (and are once again becoming popular in the West as health foods). Huts constructed with local materials, rather than indicating poverty, represent an ecologically more evolved method of providing shelter than concrete houses in many conditions. Similarly, natural fibers and local dress are far superior in satisfying region-specific needs to machine-made nylon clothing, especially in tropical climates. The West uses its own misguided definition of poverty and backwardness to legitimize nonsustainable forms of development, which have, in turn, created further conditions for material poverty or misery by diverting essential resources to resource-intensive production processes.

Once we break free of the mental prison of separation and exclusion and see how the world is interconnected, new alternatives emerge. Despair turns to hope. Violence gives way to nonviolence. Scarcity transforms into abundance and insecurity to security. Diversity becomes a solution to violence, not its cause.

The concrete context of culture—the food we eat, the clothes we wear, the languages we speak, the faiths we hold—is the source of our human identity. However, economic globalization has hijacked culture, reducing it to a consumerist monoculture of McDonald's and Coca-Cola on the one hand, and negative identities of hate on the other.

The Cartesian idea of freedom is based on separation and independence. This conception of independence has its roots in capitalist patriarchy and allows powerful men owning capital and property, while dependent on women, farmers, workers, and other cultures and species, to pretend that they are independent. Furthermore, these men can pretend that those whom they exploit and who support them, are dependent on them. Patriarchy presents women as dependent. Imperialism projects itself as a liberator—the colonized are dependent on the empire for freedom and liberation. Blindness to the role of others leads to arrogant power, domination, and violence. It is this arrogant blindness and lack of awareness that lead such undertakings as the Project for the New American Century (PNAC) to claim "that American leadership is good both for America and for the world; and that such leadership requires military strength, diplomatic energy, and commitment to moral principles."[3]

This paradigm of imperialist globalization violently imposes the monoculture of greed and consumerism on all societies and calls it "economic reform." It then externalizes the resulting insecurity and rise of narrow exclusivist identities, and calls those toxic residues of globalization culture.

The identity of the consumer in the global marketplace and the negative identity of the cultures of hate and fear have nothing to offer us with respect to the aspirations, meanings, and fulfillment that make us truly human. Negative identities are celebrated as the bonds that hold society together. "There is no closer tie than the one forged by bloodshed at the hands of a common enemy," proclaimed an article in *Business Week* after the 2004 terrorist attack in Madrid. This negative identity, forged through hate, is a distorted one.[4]

Dying cultures kill themselves, and from their negative identities unleash violence on others. The suicide bomber has become the symbol of dying cultures, testifying to the hopelessness of negatively experienced identity.

Identities can also be forged by compassion and the consciousness that we all belong to the earth family. These deep positive identities recognize that we share a common evolutionary history and a common future. They are stronger than those forged from hate. We, especially indigenous peoples, have a deep identity of place. We have bonds of family, community, and country. We have an identity as members of the earth family. We have a common human identity that is universal, even while embedded in local culture. We are both local and universal beings. Living cultures are vibrant, evolving, self-generative, and peaceful. Living cultures are rooted in life—the life of the earth, the life of the community.

The economic, ecological, and social crises resulting from corporate globalization demand a new way of thinking and living on this planet. They demand a new worldview in which compassion, not greed, is globalized; a new consciousness in which we are not reduced to consumers of globally traded commodities if we are privileged, or to narrow, fragmented one-dimensional identities based on color, religion, or ethnicity if we are excluded. We can and we do experience our lives as planetary beings with planetary consciousness, mindful and aware of what our actions, our consumption, cost other humans, other species, and future generations.

Not only are we connected with all life on the planet, past and future, but the diverse and multiple dimensions of our lives are connected. Economy shapes culture, culture shapes economy.

Earth Democracy reconnects culture to how and what we produce and consume, and to how we govern ourselves.

Beginning with people's everyday actions, Earth Democracy offers a potential for changing the way governments, intergovernment agencies, NGOs, and corporations operate. It creates a new paradigm for global governance while empowering local communities. It creates the possibility of strengthening ecological security while improving economic. security. And, on these foundations, it makes societies immune to the virus of communal hatred and fear.

Earth Democracy offers a new way of seeing, one in which everything is not at war with everything else, but through which we can cooperate to create peace, sustainability, and justice in our violent and volatile times.

Earth Democracy provides the context for living cultures—inspired both by the timeless wisdom of ancient worldviews and by the emerging solidarities of new global movements of citizens against globalization, war, and intolerance. Humanity has been connected through a planetary consciousness in the past. Our contemporary crises—the multiple fallouts of globalization—connect our future humanity even more intimately. We are experiencing ourselves as simultaneously local, national, and global. This diversity and multiplicity, and the nonviolence and inclusiveness it implies, is giving birth to a new living culture of our common humanity and our rich diversities.

As Gandhi has said, nonviolence is not just the absence of violence. It is an active engagement in compassion. *Ahimsa*, or nonviolence, is the basis of many faiths that have emerged on Indian soil. Translated into economics, nonviolence implies that our systems of production, trade, and consumption do not use up the ecological space of other species and other people. Violence is the result when our dominant economic structures and economic organization usurp and enclose the ecological space of other species or other people.

According to an ancient Indian text, the *Isho Upanishad:*

> The universe is the creation of the Supreme Power meant for the benefits of [all] creation. Each individual life form must, therefore, learn to enjoy its benefits by forming a part of the system in close relation with other species. Let not any one species encroach upon other rights.[5]

Whenever we engage in consumption or production patterns which take more than we need, we are engaging in violence. Non sustainable consumption and nonsustainable production constitute a violent economic order.

In the *Isho Upanishad* it is said:

> A selfish man overutilizing the resources of nature to satisfy his own ever increasing needs is nothing but a thief, because using resources beyond one's needs would result in the utilization of resources over which others have a right.[6]

The Eurocentric concept of property views capital as the only kind of investment and, hence, treats returns on capital investment as the only kind that need protection. Non-Western indigenous communities and cultures recognize that investments can also be of labor or of care and nurturance. Such cultural systems protect investments beyond capital. They protect the culture of conservation and the culture of caring and sharing.

Ahimsa combines justice and sustainability at a deep level. "Not taking more than you need" ensures that enough resources are left in the ecosystem for other species and the maintenance of essential ecological processes to ensure sustainability. It also ensures that enough resources are left for the livelihoods of diverse groups of people.

Not taking more than we need is also the highest expression of the precautionary principle; it ensures that we avoid harm in the absence of full knowledge of the impact of our actions.

Diversity and pluralism are necessary characteristics of an ahimsic economic order. If we don't encroach on others' rights, diverse species will survive and diverse trades and occupations will flourish. Diversity is, therefore, a barometer of nonviolence and reflects the sustainability and justice that nonviolence embodies.

Diversity is intimately linked to the possibility of self-organization. It is, therefore, the basis of both swadeshi and swaraj, of economic and political freedom. Decentralization and local democratic control are political corollaries of the cultivation of diversity. The conditions in which diverse species and communities have the freedom to self-organize and evolve according to their own needs, structures, and priorities are also conditions for peace.

Living societies, living ecosystems, living organisms, and living cultures are characterized by three principles:

1. the principle of diversity;
2. the principle of self-organization, self-regulation, and self-renewal;
3. the principle of reciprocity between systems, which is also called the law of return, the law of give and take.

Our diversity makes mutuality and a culture of give-and-take possible. Mutuality makes self-organization possible. Deeply autonomous and self-organized, yet deeply connected—with the earth, all species, and each other—humans are creating conditions for their future survival. An Earth Democracy is being reborn, even as we are surrounded by violence and war.

Globalization and Culture Wars

Economic globalization is not merely responsible for economic wars and class division. It is also contributing to cultural wars and religious and ethnic conflicts. When the monoculture of economic globalization is imposed on ethically and religiously diverse societies, the diversity is not eliminated—it mutates into virulent forms emerging as religious fundamentalism, ethnic cleansing, and other symptoms of cultural wars. These cultural mutations are induced by multiple factors.

As Amy Chua discusses in her book *World on Fire*, the economic polarization of globalization is superimposed on existing class inequalities. These class inequalities frequently mirror ethnic patterns. Class conflicts, she argues, thus get camouflaged as ethnic conflicts.

As diverse cultures experience a threat to their values, norms, and practices by globalization, there is a cultural backlash. When the cultural response does not simultaneously defend economic democracy and create living economies, it takes the form of negative identities and negative cultures.

Culture and economy are inseparable. The neoliberal ideology of development and globalization wishes culture away, yet culture dominates and becomes the surrogate for concerns over livelihoods and economic security. Fundamentalist religion becomes, as Marx so aptly observed, an "opiate of the masses."

Politicians and political parties that have fully supported the agenda of neoliberal globalization are also increasingly invoking exclusivist religion for gaining political power—and claiming their power comes directly from God, not from corporation and capital. The "divine right to rule" seems to be the epidemic of the day. A concept that died with feudalism is making a comeback through representative democracy in the context of globalization.

Pulitzer Prize–winning columnist Maureen Dowd points out how the evangelical fundamentalist Christians who brought President Bush to power a second time do not believe in Christian teachings like "love thy neighbor," "good will toward men," "blessed be the peacemakers," and "judge not lest you be judged." She quotes the evangelist Bob Jones who has written to the president that "Christ has allowed you to be his servant" so he could "leave an imprint for righteousness." Then Jones goes on to define Bush's mandate: "In your reelection, God has graciously granted America—though she does not deserve it—a reprieve from the agenda of paganism. . . . Put your agenda on the front burner and let it boil. You owe the liberals nothing. They despise you because they despise your Christ."[7] The book, *The Faith of George W. Bush*, written by Christian author Stephen Mansfield, reports Bush as saying, "I feel like God wants me to run for President. I can't explain it, but I sense my country is going to need me. Something is going to happen.... I know it won't be easy on me or my family, but God wants me to do it."[8]

Similarly in India, the BJP party president L K Advent declared, "The BJP is really the chosen instrument of the divine to take our country out of its present problems and to lofty heights of all round achievement."[9]

Imperialism is both an economic and cultural process. It is no accident that there has been an emergence of an arrogant, blinded, religious zeal to rescue the fallen, the cursed, the barbaric. Today the label of barbarism is being applied to Afghanistan, Iraq, Iran, and Syria. Two centuries ago when India was the target for imperial conquest, it was viewed as needing imperial salvation. At a time when India's manufacturing and agriculture was being destroyed, a debate raged in the British House of Commons about the need to civilize and Christianize India. William Wilberfree, who described India as "deeply sunk, in the lowest depths of moral and social wretchedness and degradation," ignored the role of imperialism in impoverishing India and creating wealth in England.[10] Wilberfree ascribed the success of the British to their "religious and moral superiority"—and then prescribed that the barriers for spreading "access to the blessings of Christian light be removed so that [India's] 'brightened land' and 'desolate hearts' could beam with heavenly truth, love and consolation."[11]

The rise of empire goes hand in hand with the imperialism of religion and culture. Both share an intolerance for diversity and the illusion of deliverance through destruction. The evangelical fundamentalists in the US today had their counterparts in another age of empire. Imperialism of religion and culture simultaneously performs two functions—it hides the roots of economic injustice and dispossession and it offers cultural colonization as a cure.

Globalization as Genocide

The wars of the empire and economic wars of globalization have morphed into one. Symbolic of this convergence is the appointment of Paul Wolfowitz, a key architect of the Iraq war and the Project for the New American Century, as president of the World Bank.

Imperialistic globalization is emerging as the worst form of genocide in our times. It is turning the vast majority of the human race into threatened species. Small farmers and peasants—two thirds of humanity—are an endangered species in the agenda of globalized, corporatized agriculture. Women—half of humanity—are also becoming a threatened species as subtle changes in societal arrangements introduce imbalance, and the patriarchal biases of traditional cultures converge with patriarchal biases of global capitalism to render women disposable.

WTO Kills Farmers

Lee Kyung Hae martyred himself while wearing a sign reading "WTO kills farmers" at the Cancun WTO ministerial to attract attention to one of the worst genocides of our times—the genocide of small farmers through the rules of globalization. His suicide is merely the most public of the tens of thousands of farmers who have been driven to kill themselves. Thirty thousand farmers have been killed by globalization policies in India over a decade. According to India's National Crime Bureau, 16,000 farmers in India committed suicide during 2004. During one six-month span in 2004, there were 1,860 suicides by farmers in the state of Andhra Pradesh alone.[12]

Farmer suicide emerged in India in 1997. The policies of corporate-driven globalized and industrialized agriculture deliberately destroy small farms, dispossess small farmers, and render them disposable.

The Indian peasantry, the largest body of surviving small farmers in the world, today faces a crisis of extinction. Two-thirds of India makes its living from the land. In this country of a billion, that has farmed this land for more than 5,000 years, the earth is the most generous employer. However, as farming is delinked from the earth, the soil, the climate, and biodiversity, and is instead linked to global corporations and global markets, and as the generosity of the earth is replaced by the greed of corporations, the viability of small farmers and small farms is destroyed. Farmer suicides are the most tragic and dramatic symptom of the crisis of survival faced by Indian peasants.

Rapid increase in indebtedness is at the root of farmers' taking their lives. Debt is a reflection of a negative economy. Two factors have transformed agriculture from a positive economy into a negative economy for peasants—the rising costs of production and the falling prices of farm commodities. Both these factors are rooted in the policies of trade liberalization and corporate globalization.

In 1998, the World Bank's structural adjustment policies forced India to open up its seed sector to global corporations like Cargill, Monsanto, and Syngenta. The global corporations changed the input economy overnight. Farm-saved seeds were replaced by corporate seeds, which need fertilizers and pesticides and cannot be saved.

Corporations prevent seed savings through patents and by engineering seeds with nonrenewable traits. As a result, poor peasants have to buy new seeds for every planting season and what was a traditionally free resource, available by putting aside a small portion of the crop, becomes a commodity. This new expense increases poverty and leads to indebtedness.

The shift from saved seed to a corporate monopoly of the seed supply also represents a shift from biodiversity to monoculture in agriculture. The district of Warangal in Andhra Pradesh used to grow diverse legumes, millets, and oilseeds. Now the imposition of cotton monocultures has led to the loss of the wealth of farmer's breeding and nature's evolution.

Monocultures and uniformity increase the risks of crop failure, as diverse seeds adapted to diverse ecosystems are replaced by the rushed introduction of uniform and often untested seeds into the market. When Monsanto first introduced Bt cotton in India in 2002, the farmers lost 1 billion rupees due to crop failure. Instead of 1,500 kilos per acre as promised by the company, the harvest was as low as 200 kilos per acre. Instead of incomes of 10,000 rupees an acre, farmers ran into losses of 6,400 rupees an acre. In the state of Bihar, when farm-saved corn seed was displaced by Monsanto's hybrid corn, the entire crop failed, creating 4 billion rupees in losses and increased poverty for desperately poor farmers.[13] Poor peasants of the South cannot survive seed monopolies. The crisis of suicides shows how the survival of small farmers is incompatible with the seed monopolies of global corporations.

The second pressure Indian farmers are facing is the dramatic fall in prices of farm produce as a result of the WTO's free trade policies. The WTO rules for trade in agriculture are, in essence, rules for dumping. They have allowed wealthy countries to increase agribusiness subsidies while preventing other countries from protecting their farmers from artificially cheap imported produce. Four hundred billion dollars in subsidies combined with the forced removal of import restrictions is a ready-made recipe for farmer suicide. Global wheat prices have dropped from \$216 a ton in 1995 to \$133 a ton in 2001; cotton prices from \$98.2 a ton in

1995 to $49.1 a ton in 2001; soya bean prices from $273 a ton in 1995 to $178 a ton. This reduction is due not to a change in productivity, but to an increase in subsidies and an increase in market monopolies controlled by a handful of agribusiness corporations.

The US government pays $193 per ton to US soya farmers, which artificially lowers the price of soya on the world market. In India, due to the removal of quotas and the lowering of tariffs, cheap soya has destroyed the livelihoods of not only soya growers but also other farmers who grow oil-producing crops, including coconut, mustard, sesame, and groundnut.

Similarly, cotton producers in the US are given a subsidy of $4 billion annually. This has artificially brought down cotton prices, allowing the US to capture world markets previously accessible to poor African countries such as Burkina Faso, Benin, and Mali. This subsidy of $230 per acre in the US is untenable for the African farmers. African cotton farmers are losing $250 million every year. That is why small African countries walked out of the Cancun negotiations, leading to the collapse of the WTO ministerial.

The rigged prices of globally traded agriculture commodities steal from poor peasants of the South. A study carried out by the Research Foundation for Science, Technology and Ecology (RFSTE) shows that due to falling farm prices, Indian peasants are losing $26 billion annually. This is a burden their poverty does not allow them to bear. As debts increase—unpayable from farm proceeds—farmers are compelled to sell a kidney or even commit suicide. Seed saving gives farmers life. Seed monopolies rob farmers of life.

The use of the word suicide obscures the social cause of this act. When viewed as the actions of individual farmers, these are suicides. When the 16,000 Indian farmer suicides in 2004 are viewed as the result of economic policy, this is not suicide; it is genocide.

A WHO report on violence has identified genocide "as a particularly heinous form of collective violence, especially since perpetrators of genocide intentionally target a population group with the aim of destroying it." The report continues, defining collective violence as:

> The instrumental use of violence by people who identify themselves as members of a group—whether this group is transitory or has a more permanent identity-against another group or set of individuals in order to achieve political, economic or social objectives.[14]

According to the UN Convention on the Prevention and Punishment of the Crime of Genocide:

> Genocide means any of the following acts committed with intent to destroy, in whole or in part, a national, ethnical, racial or religious group . . . (c) Deliberately inflicting on the group conditions of life calculated to bring about its physical· destruction in whole or in part.[15]

The WTO rules on agriculture are a deliberately designed policy to destroy small farmers and shift agriculture into the hands of agribusiness. The trade rules inflict on our small farmers conditions of life calculated to bring about their physical destruction as sovereign producers. WTO policies are, hence, a genocide on small farmers.

The policies, trade treaties, and technologies designed to open the way for corporate control of agriculture represent an instrumental use of violence against farmers. These genocidal policies are designed to push small farmers and family farms to extinction. The groups perpetrating the violence include the WTO, the World Bank and the IMP, global agribusiness corporations, and governments. The WTO's Agreement on Agriculture actually promotes "producer retirement"—a gentle phrase for the destruction of farmers' livelihoods. The World Bank, while recognizing the productivity of small farmers, noting that "small holders are out standing managers of their own resources—their land and capital, fertilizer and water," talks of the necessity of moving peasants out of what they call "subsistence" farming, that is, independent, sovereign agriculture production. In order to provide cheap labor to agribusiness and the production of artificially cheap export commodities, the World Bank recipes for development willingly rob millions of farmers of their freedom and their lives.

Globalization is leading to jobless growth, and creating "unskilled" workers by displacing skilled, knowledgeable farmers from their land and soil: The deliberate uprooting of peasants is equivalent to an absolute denial of the right to livelihood and right to life. It is an act of deliberate violence for an economic and political aim—the corporate control of the food economy.

India was among the countries that questioned the unfair rules of the WTO in agriculture and, along with Brazil and China, led the G-22 alliance. This alliance addressed the need to safeguard the livelihoods of small farmers from the injustice of free trade based on high subsidies and dumping. Yet at the domestic level, official agencies in India are in deep denial about any links between free trade and farmers' survival.

The government is desperate to delink farm suicides from the economic processes of globalization and has sought to silence discussion of this topic. An expert committee set up by the government of Karnataka recommends that: "The government should launch prosecution on the responsible persons involved in misleading the public and government by providing false information about farmers suicide as crop failure or indebtedness."[16]

However, farmer suicides cannot be delinked from indebtedness and the economic distress small farmers are facing. Indebtedness is not new. Farmers have always organized for freedom from debt. In the 19th century, the so-called Deccan Riots were farmer protests against the debt trap into which they had been pushed to supply cheap cotton to the textile mills in Britain. In the 1980s, farmers organized to fight for debt relief from public debt linked to Green Revolution inputs. Now, under globalization, farmers are losing their social, cultural, and economic identity as producers. Not just their livelihood, but their identity is under attack. Farmers, traditionally viewed as producers, are now viewed as consumer—consumers of costly seeds and costly chemicals. In the face of this inhumane, brutal, and exploitative convergence of global corporate capitalism and local feudalism, the farmer, as an individual victim, feels helpless. The bureaucratic and technocratic systems of the state are coming to the rescue of the dominant economic interests by blaming the victim.

It is necessary to stop this war against small farmers. It is necessary to rewrite the rules of trade in agriculture. It is necessary to change our paradigms of food production. Feeding humanity should not depend on the extinction of farmers and extinction of species. Another agriculture is possible and necessary—an agriculture that protects farmers' livelihoods, the earth, its biodiversity, and our public health.

Farmer suicides are a result of indebtedness, and debt is a result of rising costs of agricultural inputs and falling prices of agricultural produce. Both the rising costs of production and decline in farm prices are intended outcomes of trade liberalization and economic reform policies driven by agribusiness corporations. Farmer suicides are therefore an inevitable outcome of an agricultural policy which favors corporate welfare and ignores that of farmers. The solution to the corporate globalization-induced farm crisis offered by the government is greater corporate control over farming through contract farming. The Veeresh Committee report on farm suicides in Karnataka recommends that "Special Economic Zones for selected crops may be notified. By promoting contract farming consolidation of small and marginal holdings should be achieved and thereby such holdings should be made economically viable."[17] This prescription, however, offers the disease as the cure.

Small holdings are not economically unviable. They are in fact more productive and more efficient than large holdings. The limitation of small agricultural holdings is related to the unfairness and injustice of globalized, corporate-controlled agriculture, not size.

Productivity of small farmers is also superior, when it is measured in the context of diversity. Biodiversity-based measures of productivity show that small farmers can feed the world. Their yields represent a truly high level of productivity, composed as they are of the multiple yields of diverse species used for diverse purposes. Productivity is not lower on smaller units of land; on the contrary, it is higher. In Brazil, the productivity of a farm of up to 10 hectares was $85 per hectare while the productivity of a 500-hectare farm was $2 per hectare. In India, a farm of up to 5 acres had a productivity of 735 rupees per acre, while a 35-acre farm had a productivity of 346 rupees per acre.[18]

Deliberate destruction of small farms is therefore a policy to destroy both rural livelihoods and food security, since large industrial farms are less productive than small ones in terms of resources, energy, or nutrition.

Recommending registration of selected crops in special economic zones and the promotion of contract farming will deepen the crisis Indian peasants and Indian agriculture are facing. Growing one crop in an export zone or special economic zone means relying on a risky monoculture and the total domination by traders and corporations over prices. Such farmers are locked into dependency on traders since they do not have the cushion of diversity. They must sell what they have grown in order to survive, and they must sell at the price the buyer dictates. Contract farming is a form of serfdom, pushing the producer into deeper levels of debt and dependency.

Agri-Export Zones (AEZs) were created with the aim of facilitating the export of agricultural commodities and providing higher prices to the farmers for their agriculture produce. The aim of AEZs is to make India a strong player in the world agricultural market. Toward this end the Agriculture and Processed Food Product Export Development Authority (APEDA) has also launched a scheme known as Virtual Trade Fair

(VTF). Indian fruits, vegetables, and other agricultural products can now be bought and sold online across the globe. It is anticipated that with the development of AEZs, India's farmers will rise to the occasion and make a mark in the international agriculture trade with a "farm-to-port" approach. These zones are expected to transform select rural regions. But all of India is an agricultural economy; all regions need investment and support. AEZs will lead to an underdevelopment of agriculture in non-AEZ regions. Not only this, AEZs do not resolve the problems which lead to farmer suicides.

The highest rates of suicide are in Andhra Pradesh and Punjab, two states with the highest dependence on cash crops, the highest penetration of Monsanto's seeds, and the highest levels of corporatized agriculture. The states in which farmers are using their own seeds and growing crops for their sustenance and local markets are avoiding the debt trap that forces farmers into despair and hopelessness.

The crisis faced by potato growers shows that the government's obsession with special export zones and contract farming is part of the problem, not a solution. In Uttar Pradesh, when export zones for potatoes were created, farmer suicides began. While farmers are spending 255 rupees per quintal on production, potatoes are being sold for as little as 40 rupees per quintal, leaving farmers at a loss of 215 rupees for every quintal produced. The costs of production are between 55,000 and 65,000 rupees per hectare, of which 40,000 rupees represents the cost of seed alone. Corporations like Pepsi and McDonald's get cheaper potatoes for chips and "freedom fries," but their profits squeeze the life out of small farmers.

The crisis for potato growers, like the crisis for producers of tomato, cotton, oil seed, and other crops is directly related to World Bank—and WTO-driven trade liberalization policies. The policies of globalization and trade liberalization have created the farm crisis in general and the potato crisis in particular at three levels.

1. A shift from "food and farmer first" to "trade and corporation first" policies.

2. A shift from diversity of agriculture to monocultures and standardization, chemical and capital intensification of production, and deregulation of the input sector, especially seeds, leading to rising costs of production.

3. Deregulation of markets and withdrawal of the state from effective price regulation leading to a collapse in prices of farm commodities.

The new agriculture policies are based on withdrawing support for farmers and creating new subsidies for agro-processing industries and agribusiness. In a debate on the potato crisis, the Uttar Pradesh agriculture minister drew attention to subsidies given for cold storage and transport as a response to the crisis. These subsidies do not go to farmers. They go to traders and corporations. Pepsi's entry in Punjab highlights this trade-first policy. When the market rate of tomatoes was 2 rupees per kilo, Pepsi was paying farmers only 0.80 to 0.50 rupees per kilo and collecting 10 times that amount as a transport subsidy from the government. Cold storage owners in Uttar Pradesh have received 500 million rupees in subsidies, but these subsidies do not benefit farmers. A farmer *pays* the cold-storage owner 120 rupees per sack for storage. Cold storage owners are hiking charges to exploit the crisis. With the large scale of potato production in UP this is a massive diversion of financial resources from indebted farmers to traders, and from producers to business and industry.

While the government keeps announcing procurement-price and procurement-center gimmicks, government intervention in price regulation and procurement has all but disappeared under globalization. The government announced 195 rupees per quintal as the procurement price of potatoes, and the opening of eight procurement centers. However, no actual government procurement is being done to support farmers and ensure a fair price. Prices have therefore fallen to between 40 and 100 rupees per quintal, a bonanza for the agro-processing industry which makes even more profits from chips, but a disaster for the grower who, in despair, is being pushed to suicide. The agro-processing industry pays farmers 0.08 rupees for the potatoes that go a 200 gram bag of chips sold for 10 rupees. For 13 million metric tons of potatoes, this amounts to a transfer of 20 billion rupees from the impoverished peasants of Uttar Pradesh to global corporations.

The World Bank recipe for India is to grow vegetables for export. AEZs are part of this policy. However, India is spending nearly three times more buying vegetables from world markets than it is earning from exporting them. India sold $248 million worth of vegetables in 2002 as a result of policies focusing on vegetable exports, but it imported $678 million worth of vegetables.

The shift from farmer first to corporation first, from food first to trade first is supposed to increase Third World exports, increase farmers' incomes, and reduce Third World poverty. The opposite is happening. The export of tea declined from over 211,000 metric tons in 1997–98 to 128,000 in 2002–03. The value of tea exports dropped from 20 billion rupees to 11 billion in that period.[19] Export obsession is reducing exports and reducing incomes; instead of ending poverty it is ending farmers' lives.

Globalization of agriculture is failing at every level for the planet and people. But global agribusiness profits from this war against farmers and the land.

A nonviolent agriculture must become the core of our search for peace. It is a culture based on cocreation and coproduction with the earth. The renewal of agricultural traditions to sustain the earth and all her beings must be at the core of creating living cultures. Agriculture accounts for 70 percent of land use, 70 percent of water use, and 70 percent of the livelihoods on the planet. The seeds of these living cultures are being planted everywhere. In the midst of the devastation caused by industrialized, globalized agriculture, new food cultures are emerging. Wherever such initiatives are growing, small as they are, culture is changing.

The living cultures of food and farming bring the ecology movements, the animal rights movements, the farmers movements, and the consumer movements together in a new movement based on the protection of biological and cultural diversity. Monocultures and monopolies are giving way to diversity and coproduction. Scarcity is giving way to abundance. Insecurity is giving way to security. Agriculture as war is being replaced by agriculture as peace—peace of the soil, peace of the planets, peace of the animals, and peace of the people.

Globalization and Crimes Against Women

Globalization as a project of capitalist patriarchy has accelerated and deepened the violence against women. Globalization robs women of their productivity and creativity. Food and water, which have been provided through women's work and knowledge, are now being made into corporate commodities. And as women are displaced from productive roles in society, they are rendered disposable.

The explosion of trafficking in women is another dimension of the impact of globalization. The sex industry is often the only survival option left to women who are economic refugees in the globalized economy. In a commoditized world, women too become mere commodities to be bought and sold, traded and consumed. Trafficking in women has grown most where globalization has destroyed women's work, especially in Asia and Eastern Europe.

Women: The Providers of Food and Water

Women have been the primary producers in the sustenance economy. They are the providers of food and water, of health, and social security.

"Growth" in the global economy has led to the destruction of nature's economy—through which environmental regeneration takes place. It has also destroyed the people's sustenance economy—within which women work to sustain society. Ironically, this hard, unpaid labor is frequently denied the name of work.

When the market economy runs into trouble it is the informal economy (which consists predominantly of women's work) that must fill in the gaps and pay to restore it to health. In many cases, governments' efforts to reduce the level of a country's fiscal deficit are implemented by making substantial cuts in social and economic development expenditure and, as a result, real wages decrease considerably. In times of structural adjustment and austerity programs, cuts in public expenditure generally fall most heavily on the poor and on women.

The Trade Metaphor Versus Nature's Economy

As the trade metaphor has come to replace the metaphor of home, the meaning of *value* itself has been transformed.

Value, which means worth, has been redefined in terms of exchange and trade. Unless something is tradable it has no economic value. This assumption that something only has value if it can be exchanged for money has also rendered nature's economy worthless when, in fact, it is priceless. The marginalization of both women's work and nature's work are linked to perceptions of home as a place where nothing of economic value is produced.

The completion of the Uruguay Round of GATT, and the establishment of the WTO, on January 1, 1995, has drawn all domestic issues into the global economy. Furthermore, all matters related to life—ethics, values, ecology, food, culture, knowledge, and democracy—have been brought into the global arena as matters of international trade.

Needless to say, this has also brought the perspectives, and situations, of women in the remotest villages of the South into a direct collision with the perspectives and power of men who control global patriarchal institutions.

Braving a New Analysis

Within this period of globalization, gender analysis needs to make two major shifts. First, since globalization manifests itself primarily as a removal of national barriers to trade and investment, gender analysis needs to move beyond an exclusively domestic model of analysis (limited to either the household or country) and toward an understanding of gender relations between actors at the global level.

Second, gender analysis needs to move from a focus on the end result, which victimizes women by only concerning itself with the impact on women. In order to effect change we need to adopt a structural and transformative analysis that addresses the underlying forces that form society. Global financial trade and corporate institutions are "gendered" institutions; they impact on men and women, the rich and poor, and different peoples in different ways.

These institutions and structures are created, dominated, and controlled by men. Because they are shaped by a particular gender, class, and race of humans, predominantly men from the rich G7 countries, these institutions are expressions and vehicles of the visions, aspirations, and assumptions of this particular group.

Gender analysis of globalization, therefore, cannot limit itself to the impact on women. It needs to take into account the patriarchal basis of the paradigms, models, processes, policies, and projects advanced by these global institutions. It needs to take into account how women's concerns, priorities, and perceptions are excluded in defining the economy, and excluded from the process of defining economic problems and proposing and implementing solutions.

The Convergence of Patriarchies and Fundamentalisms

Religion and capitalism have common roots in patriarchy. They have conventionally been viewed as opposing forces with religious patriarchy defending traditions and capitalist patriarchy pushing for progress and modernity. The emergence of a world order driven by global capital—globalization—is, therefore, often seen as undermining religious patriarchy and as liberatory for women. However, what we are witnessing in contemporary times is not a contest between, but a convergence of, religious and capitalist patriarchy in the form of religious fundamentalism and market fundamentalism.

Religion is often opposed to the market. However, when the market itself becomes the ruling religion and takes on the form of market fundamentalism, it can actually converge with religious fundamentalism. The convergence occurs at multiple levels.

Both market fundamentalism and religious fundamentalism make women as human beings disappear. Women are reduced to sex objects or reproductive machines to be controlled by men—either through the market or through the invocation of religious texts.

As market fundamentalism generates economic insecurity, people move to religious fundamentalism as a source of security, reinventing identity to deal with the culture of insecurity. Right wing ideologies grow in direct proportion to the insecurities generated by deregulated markets.

Globalization erodes democratic control over economic processes. Representative democracy becomes empty of economic content. The vacuum is filled with xenophobic, exclusivist fundamentalist ideologies that are divorced from people's real needs but offer an illusion of security.

Religion, which, in its embedded, inclusive, relational form, could be a countervailing value system to the excesses of the market, has become, in its fundamentalist form, part of a vicious cycle of violence and exclusion.

Globalization has unleashed a contest between women-centered worldviews, knowledge systems, and productive systems that ensure sustenance and sharing and patriarchal systems of knowledge and the economy based on war and violence. Because the division of labor has left the sustenance economy primarily in women's hands, women generate, sustain, and regenerate life. Global patriarchal institutions are unleashing death and destruction by trying to own and commodify life. The issues are old, the instruments are new. The paradigms are old, the projects are new. The patriarchal urge to control and own everything is old, the expressions are new. The ecological and feminist struggle to protect life is ancient, the context of the globalized economy is new. The epic contest of our times is about staying alive.

When food and water are hijacked by corporations for profits, women's economies and knowledge systems are destroyed—and as the marginalization of women increases, so does violence against them.

Capitalist patriarchy and religious patriarchy share the following aspects: domination of men with religious or economic power over other humans and the earth; devaluation of women, workers, and other beings; and disconnection from the earth and living cultures and economies.

Female Feticide: Disappearing Women

The prevailing view is that economic globalization will modernize societies and improve women's status. However, the opposite seems to be happening. The patriarchal values of the market are combining with anti-women values of religious patriarchy, leading not just to women's marginalization, but to their very dispensability. The growing phenomenon of female feticide in India illustrates how capitalist patriarchy and traditions of religious patriarchy are converging to unleash new levels of violence against women. This regression is no accident.

The Green Revolution region of Punjab was the place where female feticide started.[20] This region was also the first to turn amniocentesis into a tool for female feticide. The emergence of the technologies of amniocentesis and ultrasound have enabled families to know the sex of the unborn child and selectively abort female fetuses. Between 1978 and 1983, 78,000 female fetuses were aborted after sex-determination tests. The declining child sex ratio reveals the extent of the crisis. The child sex ratio is calculated as the number of girls per 1,000 boys in the 0–6-year age group. There is a systematic decline in the child sex ratio which has dropped from 976 in 1961 to 927 in 2001. The decline has been sharpest since 1981, as sex-determination technology became increasingly available.

Child Sex Ratio, 1981–2001

Year	Sex ratio (age 0-6, girls per 1,000 boys)	Variation
1961	976	—
1971	964	–12
1981	962	–2
1991	945	–17
2001	927	–18

Discrimination has mutated into dispensability under the pressure of globalization. The traditional bias for male children has been combined with the commodification of life itself to further devalue women. The convergence of patriarchies is becoming a threat to women's very survival.

India's population grew 21 percent between 1991 and 2001 to 1.03 billion people. While the population grew, girls were disappearing. The change in sex ratio combined with population growth reveals there are 36 million fewer females in the population than would be expected. This is half the world's 60 million "missing" women—those women who were not allowed to be born because of sex-selective abortion.[21]

Globalization destroys jobs and livelihoods but creates consumerism. One source of meeting consumerist desires is the coercive extraction of dowry. The practices of many communities of bride price in which the groom's family pays the girl's family and the giving of *stree-dhan*—wealth that stays in the hands of women—are being replaced by dowry for luxury consumption. The bride price is given to a girl's family in acknowledgement that the family is losing a productive contributor to the household sustenance economy. Dowry, on the other hand, is extracted from the girls family and devalues women by defining them as a burden. The spread of dowry—used largely for purchasing consumer goods such as cars, televisions, and refrigerators—is contemporaneous and contiguous with the spread of the culture of consumerism.

The burden of the dowry is taking women's lives. The phenomenon of "dowry deaths," which has emerged over the past few decades, refers to the killing of wives by their in-laws because ever increasing demands for "dowry" from the girl's family could not be met. More than 5,000 women are victims of dowry deaths annually in India.[22] Female feticide emerges from the combination of the erosion of women's livelihoods and dowry demands.

As the menace of dowry spreads across the country, and across classes, the dispensability of the girl child also increases. About 84 percent of gynecologists currently perform amniocentesis in Bombay and view it as a "human service to women who do not want any more daughters."[23] A study of gynecologists in Bombay found that 64 percent carry out "amniocentesis tests solely for sex determination."[24] The monetary costs of a sex-determination test and abortion are lower than the thousands of rupees needed for a girl's dowry. In a world increasingly dominated by capitalist patriarchy, cash is the only measure of worth—of women as of everything else.

Despite the passage of the Pre-Natal Diagnostic Techniques (Regulation and Prevention of Misuse) Act of 1994, which prohibit the practice, new reproductive technologies are being increasingly used for femicide.

> Abortion has become a suicidal substitute for getting rid of unwanted pregnancies. Not only the female fetuses are being destroyed but that women resort to repeated abortions, which are indicative of their poor control over their own bodies, and their inability to exercise their right to safe sex. To add insult to injury, sex selective abortions are becoming rampant in India especially in its relatively more affluent highly gender discriminatory north-western region with acute son preference.[25]

Vibhuti Patel, a professor at Mumbai's SNDT women's university, reflects on an advertisement encouraging amniocentesis as a tool for sex-selective abortions reading:

> "Better Rs. 5000 now than Rs. 5 lakhs later" i.e. better spend Rs. 5000 for female feticide than [500,000 rupees] as dowry for a grown up daughter. By this logic, it is better to kill poor people or third world masses rather than let them suffer in poverty and deprivation. This logic also presumes that social evils like dowry are God-given and that we cannot do anything about them. Hence victimize the victim. Investing in daughter's education, health, and dignified life to make her self dependent are far more humane and realistic ways than brutalizing pregnant mother and her would be daughter.[26]

The introduction of industrial, chemical agriculture in Punjab displaced women from rural livelihoods. While women's work accounts for a major share of agricultural production where industrial farming has not robbed women of livelihoods, in Punjab's fields one sees men driving tractors and spraying chemicals. First women disappeared from productive roles in farming. Now they are disappearing from society through female feticide. The devaluation of women and the emergence of new technologies combine to pronounce a death sentence on unborn girls.

If female feticide were only a result of a traditional bias against women, it would be restricted to areas where the bias against girls has been extreme in the past and it would decline as socioeconomic changes erode traditional structures. However, female feticide is spreading like a plague across Indian society: The regions with high economic growth and more rapid "modernization" and integration with the global economy exhibit higher rates of female feticide and lower child sex ratios. As Kamalesh, a woman activist from the Haryana region, reports, things have never been so bad: "Earlier there was a guilt about it. Now it is seen as a matter of choice."[27] The greater the economic growth and prosperity the larger the number of missing girls. In Punjab, Haryana, Delhi, and Gujarat—the prosperous northwest—the ratio has declined to less than 900 girls per 1,000 boys. It has fallen to 770 in Kurukshetra, 814 in Ahmedabad, and 845 in southwest Delhi.[28] States with lower economic growth and global economic integration such as Kerala, Goa, Sikkim, Mizoram, and Tripura have not registered a decline in child sex ratio.

The report "Missing Girls: A Case Study of Delhi" from the Ministry of Health and Family Welfare states:

> At 1991 Census count more than 71,000 female infants and children below six years were missing. The 2001 Census brings out the stark fact of Delhi decimating its born and unborn female children. There are now 139,173 fewer females in the age group 0–6 years. The city continues to decimate females at inception, in the womb and currently before conception. The alarm bells were ringing in 1991 and the 2001 census sounds the death knell for the female half of our population.[29]

Delhi, India's capital region, has the highest literacy rates, the highest per-capita income, and also the most violence against women. From 1991 to 2000 overall literacy went up from 75 percent to 82 percent, while female literacy increased from 67 percent to 75 percent. However, while opportunities in the marketplace have increased, the hazards to women have also increased. The more a region is globalized, the more violence there is against women. Delhi accounts for 32.9 percent of rape, 23.3 percent of molestation, and 17.4 percent of dowry deaths—the highest rate among the 23 mega cities in India.[30]

The convergence of many forces are denying women their very right to life. The displacement of women from productive work, the destruction of economies of sustenance, and the growth of the culture of consumerism and commodification, have all combined to devalue women in society. The emergence of sex-detection technologies and increasing incomes to pay for their use has reinforced patriarchal biases instead of weakening it. Sex selection through abortion of the female fetus is becoming the ultimate tool for the dispensability of women.[31]

Women as Guardians and Promoters of Life-Centered Cultures

While capitalist and religious patriarchy and market and religious fundamentalism converge to threaten life on earth, women are responding with nonviolence and compassion to defend life on earth and resist violence against women.

Globalization is redefining the very status and understanding of creation. Corporations like Monsanto pirate and patent the qualities of ancient Indian wheat and become "creators" and "inventors" of plants. Corporations like Suez and Vivendi and Coca-Cola refer to water in rivers and in underground rivers as "raw" material, and the water they package and sell as their product. Capitalist patriarchy thus defines creation and nature as raw material, and acts of domination, destruction, and exploitation acts of creation. In science and technology, this patriarchal myth of creation has led to patents on life and laws like the TRIPS agreement of WTO. In the economic sphere, patriarchal values have led to destruction of household and local economies, and of entire ecosystems. The destruction is counted as "growth." Dispensability is interpreted as liberation.

What happened in the Iraq war was destruction. It is being referred to as reconstruction. Innocent people were killed; thousands of years of the history of civilization was destroyed and erased, yet Jay Garner—the retired US general unilaterally appointed as head of the Office for Reconstruction and Humanitarian Assistance—talked about giving birth to a new system in Iraq.

Bombs do not give "birth" to society. They annihilate life. New societies are not "born" by destroying the historical and cultural legacy of ancient civilizations. Maybe the choice to allow the destruction of Iraq's historical legacy was a prerequisite for this illusion of giving birth to a new society.

Maybe the rulers in the US do not perceive these violations because their own society was built on the genocide of Native Americans. Annihilation of the other seems to be taken as natural by those controlling power in the world's lone superpower. Maybe the perception of the deliberate destruction of a civilization and thousands of innocent lives as a birth process is an expression of the Western patriarchy's illusion of creation. This illusion confuses destruction with creation and annihilation with birth. This illusion portrays capital and machines, including war machines, as sources of creation and nature and human societies, especially non-Western societies, as either dead, inert, and passive or dangerous and cannibalistic. This worldview creates the "white man's burden" for liberating nature and our societies and sees that "liberation" as the birth of freedom.

Religions that recognize the integrity of creation and the sanctity of life are a source of resistance to this destruction. And while men in power redefine religion in fundamentalist terms and in support of market fundamentalism, women in diverse cultures mobilize their faith, their spirituality, their power to protect the earth, and life on earth. Despite being subjected to the double burden of religious and capitalist patriarchy, women are emerging as leaders and guardians of life-centered cultures, economies, and policies. Movements to defend water are being led by women. Movements to defend biodiversity are being led by women. Movements for food rights are being led by women. While overcoming their marginalization, women are emerging as guardians of life and the future.

Women farmers have kept and bred seeds over millennia. Basmati is just one among 100,000 varieties of rice evolved by Indian farmers. Diversity and perenniality are principles that form the foundation of our culture of the seed. In central India, at the beginning of the agricultural season, farmers gather at the village deity, offer their rice varieties, and then share the seeds. This annual festival of Akti rejuvenates the duty of saving and sharing seed among farming communities. It establishes a partnership among farmers and with the earth.

Gender inequality and women's exclusion and dispensability arise from systems of patriarchy organized through religion or economic and political systems. Separation from life and living processes enables anti-life systems to become dominant. Gender equity requires seeing women in their full humanity—as producers and creators, as custodians of culture, as political decision-makers, as spiritual beings. Women's full humanity becomes the healing force that can break the vicious cycles of violence based on treating the inhumanity of man as the measure of being human, of greed as the organizing principle of the economy, of genocide and suicide as the expression of religious fervor. Underlying the exclusivist philosophies of the fundamentalist market ideology and religious ideology is the polarization of identity. In Indian philosophy, we think in terms of *so-hum:* "You are, therefore I am." Fundamentalisms, however, function on the belief "If you are, I am not," or "My existence requires your annihilation." Samuel Huntington's *Clash of Civilizations* is based on this paradigm of mutual exclusion, hence mutual annihilation: "For peoples seeking identity and reinventing ethnicity, enemies are essential."[32]

Women are refusing to be part of the culture of hate and violence. Women, in and through their lives, are showing that love and compassion, sharing and giving are not just *possible* human qualities—they are necessary qualities for us to be human.

Patriarchies have modeled "being human" on inhuman, violent, acquisitive, exploitative, and destructive traits. Women are redefining being human as a capacity to care and to share, to love and protect, to be guardians not owners of nature's gifts, and to find strength and security in diversity, not in oppressive monocultures. What patriarchy defined as a source of weakness is in fact the source of strength. And the pseudo-strengths of patriarchy based on violence and domination have their roots in weakness that comes from separation and alienation, from fear of others and others' freedoms and sovereignty. We are witnessing the worst of violence in our times. But we are also reinventing nonviolence and finding new courage to defend life, including our own.

In 1997, Jean Grosshltz of Mt. Holyoke College, Christine von Weisacker of Germany, Beth Burrows of the Edmonds Institute, and I decided to come together to form "Diverse Women for Diversity" to reclaim the future of all beings, we declared:

> We women, in all our vibrant and fabulous diversity, have witnessed the increasing aggression against the human spirit, human mind, and human body and the continued invasion of, and assault upon the Earth and all her diverse species.
> We demand of governments, international organizations, transnational corporations, and individual men who share our rage that they address the

> crisis that has been caused by the creation of monocultures and the reduction, enclosure, and extinction of biological and cultural diversity.
> We must insist that those who would address the crisis listen to and take leadership from women, indigenous peoples, farmers, and all who have raised these concerns at the local level. We ask them to hear those whose wisdom, stewardship, knowledge, and commitment has already been demonstrated by the preservation of the diversity we celebrate today.

Globalization has two meanings. It can refer to our universal humanity, to cultures of compassion and solidarity, to our common identity as earth citizens. I call this globalization Earth Democracy. The dominant meaning and form of globalization is economic or corporate globalization. This is the globalization of capitalist patriarchy—in which everything is a commodity, everything is for sale, and the only value a thing has is the price it can bring in the global marketplace. All other values are merely "tariff and non-tariff" trade barriers. Noncommercial values, and the lives and cultures they support, are excluded.

As Walt Martin and Magde Ott write in their book *Albert Schweitzer: Reverence for Life:*

> Unity in diversity, diversity in unity is the natural state of all life. We are interdependent—plants, animals, people—and need the delicate interplay between these various expressions of life to remain healthy. This natural interdependence of human kind is not the artificial "globalization" forced on the many by a few. It comes from the people, is organic, basic to our lives, and originates from the profound experience that life is universal.

We want to live, and "live in the midst of other life that wants to live" as Schweitzer succinctly put it.[33]

This irrepressible urge to live and celebrate life in its diversity is the basis of creating living cultures. Living cultures nourish life; they do not spread death and destruction, fear and insecurity. Living cultures evolve from our connectedness with all life. Cultures are based on identity. However, corporate globalization and fundamentalism reduce and manipulate our identities. As part of the universe, we have a universal identity. As part of the earth family, we are earth citizens and have earth identities which are both the particular identity of place, and the global planetary identity. As part of countries we have citizenship. As members of communities, we have multiple community identities—what we do, what we eat, what we wear, what we speak. These diverse, multiple identities shape our sense of self and who we are. And these diversities are not inconsistent with our common humanity. Without diversity, we have no humanity.

Living cultures are cultures of life, based on reverence for all life—women and men, rich and poor, white and black, Christian and Muslim, human and nonhuman.

Reverence for life is based on compassion and caring for the other, recognition of the autonomy and subjecthood of the other, and the awareness that we are mutually dependent on each other for sustenance, for peace, for joy.

When millions around the world marched for peace on February 15, 2003, they marched for solidarity, not domination, as the basis of human security and freedom.

Freedom in living cultures is based on interconnection and interdependence. The Declaration of Interdependence sponsored by the Democracy Collaborative is an expression of the emergent movement for peace that is based on protecting our diversities and our human commonality.

Declaration of Interdependence

We the people of the world do herewith declare our interdependence as individuals and members of distinct communities and nations. We do pledge ourselves citizens of our CivWorld, civic, civil, and civilized. Without prejudice to the goods and interests of our national and regional identities, we recognize our responsibilities to the common goods and liberties of humankind as a whole.

We do therefore pledge to work both directly and through the nations and communities of which we are also citizens:

- To guarantee justice and equality for all by establishing on a firm basis the human rights of every person on the planet, ensuring that the least among us may enjoy the same liberties as the prominent and the powerful;

- To forge a safe and sustainable global environment for all—which is the condition of human survival—at a cost to peoples based on their current share in the world's wealth;

- To offer children, our common human future, special attention and protection in distributing our common goods, above all those upon which health and education depend;

- To establish democratic forms of global civil and legal governance through which our common rights can be secured and our common ends realized;
- To foster democratic policies and institutions expressing and protecting our human commonality; and at the same time,
- To nurture free spaces in which our distinctive religious, ethnic and cultural identities may flourish and our equally worthy lives may be lived in dignity, protected from political, economic and cultural hegemony of every kind.[34]

CHAPTER FOUR

Earth Democracy in Action

Over the past three decades, in the defense of my extended earth family, I have worked as an earth citizen to shape an Earth Democracy. I have tried to combine knowledge with action; I have strived to connect the local to the global. By transcending divisions, we are collectively creating new possibilities and engendering a post globalization world. I have chosen to dedicate my energies to realizing Earth Democracy in spheres vital to survival. That is why I focus on seed, food, and water. Through Earth Democracy in action we are reclaiming the freedoms and rights of all people and all beings. Through everyday actions on everyday issues, we are creating living economies, living democracies, and living cultures. Diversity, alliances, cooperation, and persistence are our strengths. Service, support, and solidarity are our means. Justice, human freedom, dignity, and ecological survival are our ends. We are reclaiming a world precariously on the edge. We take action not with arrogance and certainty, but with humility and uncertainty. It is our giving that counts—not our success. But in selfless giving, we have victories. And through everyday actions, we reweave the web of life.

Bija Swaraj: Reclaiming the Democracy of All Life

On March 8, 2005, International Women's Day, we won a major victory in a biopiracy case after a 10-year legal battle in the European Patent Office. The United States Department of Agriculture and W. R. Grace jointly claimed to have "invented" the use of the neem tree (*Azadirichta indica*) for controlling pests and diseases in agriculture. On the basis of this claim they were granted patent number 436257 by the European Patent Office.

Neem, or *azad darakht* to use its Persian name, which translates as free tree, has been used as a natural pesticide and medicine in India for over 2,000 years. As a response to the 1984 disaster at the Union Carbide's pesticide plant in Bhopal, I started a campaign with the slogan: "no more Bhopals, plant a neem." A decade later we found that because W. R. Grace was claiming to have invented the use of neem, the free tree was no longer going to be freely accessible to us. We launched a challenge to the neem biopiracy and more than 100,000 people joined the campaign. Another decade later, success—the European Patent Office revoked the patent.

Our success in defeating the claims of the US government and US corporations to traditional knowledge and biodiversity came because we combined research with action, and we mobilized and built movements at the local level. Three women working in global solidarity—Magda Aelvoet, former president of the Greens in the European Parliament; Linda Bullard, the president of the International Federation of Organic Agriculture Movements (IFOAM); and myself—saw the case through for over a decade without losing hope. Our lawyer, Dr. Dolder, a professor of intellectual property at Basel University, gave his best without expecting typical patent lawyer fees.

The neem victory throws light on one of the most pernicious aspects of the current rules of globalization—the WTO's Trade Related Aspects of Intellectual Property Rights agreement. TRIPS allow global corporations to patent anything and everything-life forms, seeds, plants, medicines, and traditional knowledge. Patents are supposed to satisfy three criteria: novelty, nonobviousness, and utility. "Novelty" requires that the invention not be part of "prior art" or existing knowledge; "nonobviousness" requires that someone familiar in the art would not take the same step. Most patents based on the appropriation of indigenous knowledge violate these criteria, because they range from direct piracy to minor tinkering involving steps obvious to anyone trained in the techniques and disciplines involved. Since a patent is an exclusive right granted for an invention, patents on life and traditional knowledge are twice as harmful and add insult to injury. Such patents are not based on inventions; they serve as instruments for preventing the poor from satisfying their own needs and using their own biodiversity and their own knowledge.

Patents on seeds not only allow monopolies on genetically engineered seed, they allow patenting of traditional varieties and properties used by farmers over millennia. This biopiracy is illustrated in the cases of rice and wheat.

Basmati Biopiracy

The Indian subcontinent is the biggest producer and exporter of superfine aromatic rice: basmati. India grows 650,000 tons of basmati annually. Basmati covers 10 to 15 percent of the land area under rice cultivation in India. Basmati and non-basmati rice are exported to more than 80 countries across the world. Basmati exports were 488,700 tons and amounted $280 million. Non-basmati rice exports in 1996–97 were 1.9 million tons and amounted to $450 million. The main importers of Indian basmati are the Middle East (65 percent), Europe (20 percent) and the US (10 to 15 percent). Fetching $850 a ton in the European Union (EU) compared with $700 a ton for Pakistani basmati and $500 a ton for Thai fragrant rice, Indian basmati is the most expensive rice being imported by the EU.[1]

Basmati has been grown for centuries on the subcontinent, as is evident from ancient texts, folklore, and poetry. One of the earliest references to basmati is made in the famous epic of *Heer Ranjha,* written by the poet Varis Shah in 1766.[2] This naturally perfumed variety of rice has been treasured and possessively guarded by nobles, and eagerly coveted by foreigners. It has evolved over centuries of observation, experimentation, and selection by farmers who have developed numerous varieties of the rice to meet various ecological conditions, cooking needs, and tastes. There are 27 documented varieties of basmati grown in India. The superior qualities of basmati must predominantly be attributed to the contributions of the subcontinent's farmers.

On September 2, 1997, Texas-based RiceTec was granted patent number 5663484 on basmati rice lines and grains. The patent of this "invention" is exceptionally broad and includes 20 claims within it. The patent covered the genetic lines of basmati and includes genes from the varieties developed by farmers. It thus automatically covered farmers' varieties and allowed RiceTec to collect royalties from farmers growing varieties developed by them and their forefathers.

RiceTec's strain, trading under brand names such as Kasmati, Texmati, and Jasmati, possess the same qualities-long grain, distinct aroma, high-yield, and semi-dwarf-as our traditional Indian varieties. RiceTec is essentially derived from basmati; it cannot be claimed as "novel" and therefore should not be patentable. Through a four-year-long campaign, we overturned most of RiceTec's patent claims to basmati.

Monsanto's Biopiracy of Indian Wheat

Wheat forms an integral part of the life of most Indians. It has been the principal crops in several regions of India for thousands of years. India is the second-largest producer of wheat (73.5 million tons) after China. Twenty-five-million hectares of wheat are cultivated in India. In addition to being the staple food of most Indians, wheat is closely associated with religious ceremonies and festivals. Each traditional variety has its own religious or cultural significance. The different varieties of wheat, the use of different wheat preparations in rituals, and the medicinal· and therapeutic properties of wheat have all been documented in ancient Indian text and scriptures.

Monsanto's patent registered with the European Patent Office claims to have "invented" wheat plants derived from a traditional Indian variety and products made from the soft milling traits that the traditional Indian wheat provides. Monsanto's patent claims its plants were derived from varieties of traditional Indian wheat called Nap Hal. There is no traditional Indian wheat called Nap Hal. In Hindi the word would mean "that which gives no fruit" and could be a name for Monsanto's terminator seeds. "Nap Hal" is evidently a distortion of "Nepal," since the wheat varieties were collected from near the Nepal border.

In February 2004, the Research Foundation and Greenpeace filed a legal challenge against Monsanto's biopiracy. By September, Monsanto's patent had been revoked. These victories do not mean our work is over. Corporations continue to patent life forms and pirate traditional knowledge. They also continue to impose unjust and immoral seed and patent laws on countries.

New intellectual property right (IPR) laws are creating monopolies over seeds and plant genetic resources. Seed saving and seed exchange—basic freedoms of farmers are being redefined. There are many examples of how seed acts in various countries and the introduction of IPRs prevent farmers from engaging in their own seed production. For example, Josef Albrecht, an organic farmer in Germany, was not satisfied with the commercially available seed. He worked and developed his own ecological varieties of wheat. Ten other organic farmers from neighboring villages took his wheat seeds. Albrecht was fined by his government because he traded in uncertified seed. He has challenged the penalty and Germany's seed law because he feels it restricts him from freely exercising his occupation as an organic farmer.

In Scotland, there are a large number of farmers who grow seed potato and sell seed potato to other farmers. They could, until the early 1990s, freely sell the reproductive material to other seed potato growers, to merchants, or to farmers. In the 1990s, holders of plant breeders' rights made the selling of seed potato by farmers to other farmers illegal. The seed industry began to specify the variety and the price at which they would take back the crop and barred growers from selling the crop to anyone. Soon, they started to reduce the acreage and prices. In 1994, seed potato bought from Scottish farmers was sold for more than double that price to English farmers, while the two sets of farmers were prevented from dealing directly with each other. Seed potato growers signed a petition complaining about the stranglehold of a few companies and started to sell uncertified seed directly to English farmers. The seed industry claimed they were losing 4 million pounds in seed sales because of the direct sale of uncertified seed potato between farmers. In February 1995, the British Society for Plant Breeders decided to proceed with a high-profile court case against a farmer who had been selling seeds. The farmer was forced to pay 30,000 pounds as compensation to cover lost royalties. Existing United Kingdom and European Union laws thus prevent farmers from exchanging uncertified seed as well as protected varieties.

In the US as well, farmer-to-farmer seed exchange has been made illegal. For many years, Dennis and Becky Winterboer had been deriving a sizable portion of their income from their 500-acre farm "brown bagging" their crops to other farmers to use as seed. Asgrow (a commercial company which has plant-variety protection for its soybean seeds) filed suit against the Winterboers on the grounds that its property rights were being violated. The Winterboers argued that they had acted within the law since according to the Plant Variety Act farmers had the right to sell seed, provided both the buyer and seller were farmers. Subsequently, in 1994, the Plant Variety Act was amended to eliminate farmers' privilege to save and exchange seed and establish an absolute monopoly for the seed industry.

Iraqi Order 81, promulgated and signed April 26, 2004 by Paul Bremmer imposes plant and seed patents on Iraqi farmers.[3] In 2002, FAO estimated that 97 percent of Iraqi farmers used farm-saved seed. With the new law, seed saving will be illegal.

Similar laws are being introduced in India. The entire country is being taken for a ride with claims that the Seed Act of 2004 is needed to guarantee seed quality. However, the Seed Act of 1966 already performs the function of seed testing and seed certification. Under pressure from the World Bank, India instituted a new seed policy that started to dismantle our robust public-sector seed supply system, which accounted for 20 percent of the seeds farmers use. Despite protests, the government introduced clauses on patents on life and patents on seeds in our patent law. Simultaneously the Seed Act of 2004 was introduced which would, in effect, prevent farmers from using traditional varieties. Those laws could establish a seed dictatorship, forever destroying the biodiversity of our seeds and crops and robbing farmers of all freedoms.

Eighty percent of all seed in India is still saved by farmers. Farmers' indigenous varieties are the basis of our ecological and food security. Coastal farmers have evolved salt-resistant varieties. Bihar and Bengal farmers have evolved flood-resistant varieties, farmers of Rajasthan and the semi-arid Deccan have evolved drought-resistant varieties, Himalayan farmers have evolved frost-resistant varieties. Pulses, millets, oilseeds, rice, wheat, and vegetables provide the diverse basis for our health and nutritional security. The Seed Act targets indigenous farmers' diverse varieties of crops. The Seed Act is designed to "enclose" the free economy of farmers' seed varieties." Once farmers' seed supply is destroyed through compulsory registration that makes it illegal to plant unlicensed varieties, farmers are pushed into dependency on the corporate monopoly of patented seed. The Seed Act is the handmaiden of the Patent Amendment Acts, which have introduced patents on seed.

When corporations and governments are joined in intimate nexus, noncooperation and civil disobedience are the only means for defending our freedom. Since 1994, we have been pledging that if patents on seed were introduced, we would be forced to disobey. Just as Gandhi marched to Dandi to violate the Salt Laws of the British we have started a *bija satyagraha* (civil disobedience against patents on seeds). Farmers throughout the country served notices at district offices throughout the country and more than 5 million peasants have taken a pledge to continue to save and exchange seed, and to disobey any law that prevents them from doing so.

Anna Swaraj (Food Democracy)

The globalized food system controlled and driven by agribusiness corporations is creating a fourfold crisis. The first is the crisis of nonsustainability because of overexploitation of soil and water, destruction of biodiversity, and the spread of toxic pollution from pesticides and chemical fertilizers. The second is the crisis faced by small farmers and producers. The third is the crisis of hunger, with a billion people denied their rightful shared of the earth's produce. The fourth is the obesity crisis, of which one billion people are victims, and 400,000 die annually due to diseases related to obesity. These four dimensions of our food crisis are actually caused by industrialization of food production and globalization of food distribution. This crisis is simultaneously ecological, economic, cultural, and political.

Humanity has shaped the ecology of our planet for 10,000 years. The last half century has been an experiment with nonsustainable, chemical intensive, water intensive, and capital intensive industrial agriculture. Corporate globalization is leading to food fascism—threatening the freedom of farmers and consumers and destroying the ecological, economic, and cultural foundation of food and agriculture.

The assault of corporations greedy to extract superprofits from every link of the food chain—from seed, to production, to processing and distribution—is being resisted by new movements for food democracy and food sovereignty. *Bija swaraj* (seed sovereignty) is the first link of the food chain in the movement for freedom. *Anna swaraj* (food sovereignty) is being shaped at both the production and consumption levels. Food democracy is being evolved from the local to the global level. New solidarities, new convergences are reopening spaces for food freedom, cracking the walls of dictator ship that corporations have created using the WTO, the World Bank, and national governments.

Food democracy and food sovereignty addresses all the dimensions of the crisis. By taking back control over our food systems, we can produce more food while using fewer resources, improve farmers' incomes and strengthen their livelihoods, while solving the problem of hunger and obesity. The future is not certain, but this much is, a better agriculture is possible than the one corporations offer.

The Law for Food Fascism

With the industrialization and globalization of food, food safety is a growing concern. On a global scale, new diseases and more virulent forms of old diseases are emerging as globalization spreads factory farming and industrial agriculture. Disease, epidemics, and food hazards are the outcome of production processes based on hazardous inputs and processes.

Food production technologies have undergone two generations of change over the last few decades. The first shift was the introduction of chemicals in agriculture under the banner of the Green Revolution. Toxic chemicals used in warfare were deployed in agriculture in times of peace as synthetic fertilizers and pesticides. Agriculture and food production became dependent on "weapons of mass destruction." The Bhopal disaster, in which a leak from a pesticide plant killed thousands in 1984, and has killed nearly 30,000 since then, is the most tragic reminder of how agriculture has become dependent on war technologies designed to kill.

Genetic engineering will introduce new food hazards. In the UK, more than 2 million cows were found to be infected with bovine spongiform encephalopathy (BSE) —mad cow disease. By August 2002, the death toll from variant Creutzfeldt-Jakob disease (VCJD)—the human equivalent of BSE—had reached 133. The swine fever in Asia led to the killing of millions of pigs. A newly emerged strain of the Nipah virus killed 100 pig farm workers, infected 150 with nonfatal encephalitis, and led to the slaughter of a million pigs to control the disease. The avian flu has already led to human deaths and the killing of millions of ducks and chickens. The epidemic has spread to 10 countries. The disease has jumped from chickens to humans and killed eight people in Vietnam and Thailand. The laws developed in response to food hazards are being shaped to promote large-scale globalized production, and act against local foods. These aims are also the basis of the Sanitary and Phyto Sanitary Agreement of the WTO.

In India, in August 1998, on the grounds of food safety, a new packaging regulation was introduced for edible oils that banned mustard oil and the sale of unpackaged edible oils. This law, in combination with the WTO-mandated removal of import restrictions, flooded India's markets with oil from genetically engineered soyabeans. India has used the coconut, groundnut, linseed, mustard, sunflower, and sesame for edible oil. The main consequence of eliminating import restrictions was the destruction of our oilseed biodiversity and the diversity of our edible oils and food cultures.[53]

It is also a destruction of economic democracy and economic freedom to produce oils locally, according to locally available resources, and locally appropriate food culture. Since indigenous oilseeds are high in oil content, they can be processed at the household or community level, with eco-friendly, decentralized, and democratic technologies.

The entire process of making oil from soyabeans is controlled by corporations. Monsanto controls seeds through its patents and its ownership of seed corporations. Cargill, Continental, and other trading giants control the trade and milling operations internationally. Because of the soyabean's low oil content, extraction requires heavy processing that is environmentally unfriendly and unhealthy.

Mustard oil and our indigenous oilseeds symbolize freedom for nature, for our farmers, our diverse food cultures, and the rights of poor consumers. Soyabean oil symbolizes the concentration of power and the colonization of nature, cultures, farmers, and consumers; the manipulation of oil prices and the restrictions put on indigenous oilseed processing and sales force Indians to consume soyabean oil; further strengthening a monoculture and monopoly system. In the future, with patent laws, small farmers will be forced to pay royalties for seeds and will be further pushed into poverty.

Free trade and economic globalization has been projected as economic freedom for all. However, as the case of the mustard oil crisis and soyabean imports reveals, so-called free trade is based on the destruction of economic freedom of small producers, processors, and poor consumers. Furthermore, while the rhetoric of free trade is that government should step out of business, the decision to allow free import of soyabeans and passing the packaging regulation reveals how the government is a major player in the transfer of production from small-scale decentralized systems to large-scale, centralized systems under monopoly control.

The state, in fact, is the backbone of the free trade order. Instead of regulating big business, it leaves big business free, and declares small producers and diverse cultures illegal. The asymmetric treatment of the small and the big is also evident in the regulation of food safety. While the government reacted immediately to ban mustard oil, it has done nothing to prevent the dumping of toxic, genetically engineered soyabeans. The various forms of adulteration committed by the global players is protected rather than punished, by governments in India, in the US, and across the world.

The highest political and economic conflicts—between freedom and slavery, democracy and dictatorship, diversity and monoculture—are thus waged with the simple acts of buying edible oils and cooking our food. Will the future of India's edible-oil culture be based on mustard and other edible oilseeds or will it become part of the globalized monoculture of soyabean with its associated but hidden hazards.

The EU moratorium on genetically modified (GM) crops and foods represents the will of its people not to be force-fed, and serves as an indicator of what Earth Democracy can achieve. Because of popular demand, Europe suspended its approvals process on GM crops in 1998. This action resulted in the following successes:

The indirect effects of growing GM herbicide tolerant (HT) crops on farmland wildlife was taken into account. GM HT sugar/fodder beet and spring oilseed rape, now known to be damaging to farmland wildlife, were prevented from being grown commercially in Europe.

Requirements for monitoring of environmental or human health effects were introduced, disproving agribusiness' "no evidence of harm" claim.

Consumers are able to make a choice not to eat products derived from GM crops because of the new labeling laws.

There are traceability requirements for GM foods. If adverse effects emerge, it is possible to withdraw the product from the market quickly and easily. Following BSE, traceability is a cornerstone of European food safety systems.

On May 13, 2003, the US, together with Canada and Argentina, challenged Europe's moratorium on GM crops and foods in a case still pending. Arguing that their GM products were being unfairly discriminated against, they challenged the application of the precautionary principle to decision-making about GM crops. Bringing this case to the WTO is another excuse to attack the use of the precautionary principle that is supposed to be embodied into European decision making.

The EU regulations take account of the EU's international trade commitments and of the requirements of the Cartagena Protocol on Biosafety, with respect to obligations of importers. The EU's regulatory system for GMO authorization is in line with WTO rules: It is clear, transparent, and nondiscriminatory. There is, therefore, no issue that the WTO needs to examine.

Many countries are now looking at the EU policy to develop their own policies. The US fears that other countries will adopt approaches to regulating GMOs and GM food and feed products similar to the EU's. A good example is the Swiss GM legislation that took effect January 1, 2004. It is stricter than current EU legislation on the liability and coexistence aspects. This legislation, based on the precautionary principle and the "polluter pays" principle, aims to protect the health and security of human beings, animals, and the environment. It also aims to maintain biological diversity and fertility of the soil and to allow freedom of choice for consumers.

Even with the EU moratorium on GMOs, there have been attacks in Europe on food sovereignty and attempts to implement food safety laws threatening to small producers of traditional foods. For example, a campaign by the Slow Food movement was needed to force the Italian government to amend a law that would have forced even the smallest food maker to conform to the pseudohygienic standards that suit corporations like Kraft Foods.

Food Laws and Safety for India's Diverse and Local Food Economy

Local, natural, organically grown and processed food is not the same as chemically processed food (which is different from genetically engineered food). Different foods have different safety risks and need different safety laws and different systems of management; that is why in Europe there are different standards for organic, for industrial, and for genetically engineered foods. Organic standards are set by organic movements, while the standards for genetic engineering are set at the European level through the novel food laws. There is, in addition, the movement to protect the cultural diversity of food, through "unique" and "typical" food standards, which is being destroyed by applying industrial food processing standards to locally and nonindustrially grown foods. Standards for these foods should be based on indigenous science and community control, not industrial "science" controlled by central governments and manipulated by food giants such as Cargill, ConAgra, Lever, Nestle, and Phillip Morris, and gene giants such as Monsanto.

India, like Europe, needs different laws governed at different levels for different food systems:

- An organic processing law for local, natural, small-scale food processing governed by *gramsabhas, panchayats,* and local communities. (The gramsabha is the village community as a whole when it meets for decision making; gram means village and sabha means gathering.) In cities this could be based on licensing by resident welfare associations, urban panchayats, and local municipalities. Community control through citizen participation is the real guarantee for safety.

- An industrial-processing law—which already exists in the Prevention of Food Adulteration Act. This law could be updated to deal with new food hazards.

- A GM food law which controls imports, labeling, segregation, traceability, and so forth. This is the new law that the consumers need. This law should be drafted by the central government, but states and local communities should be free to introduce stricter standards. If regions want to be GMO-free, this should be allowed under the principles of decentralized democracy.

We cannot allow integrated food safety laws to criminalize every tiny dhabawala and street vendor. Such legislation will unleash the worst form of license·and inspector raj and will establish a food fascism based on a food mafia serving global corporations. It will destroy our food freedom, our livelihoods, our food safety, and our food diversity. Ninety-nine percent of India's food is processed locally for local consumption and sale. Our science of food is based on Ayurveda, not the reductionist science that has treated unhealthy food as safe. The central government should not try to license the last dhaba in India. They are not introducing obesity, diabetes, cancer, or heart disease into our society. They are providing safe, affordable dal and roti to millions of working people.

Different food systems need different levels of management for safety. It is inappropriate to lump together all kinds of food—organic, industrial, GMOs—into one category. How food is processed determines its quality, nutrition, and safety. Home-processed bread is not the same as industrial bread. They are not "like products" to use WTO terminology. They are different products in terms of their ecological content and public health impact. A factory-raised chicken is not the same as a free-range chicken, both in terms of animal welfare and in terms of food quality and safety. GM corn is not the same as organic corn. The former contains antibiotic resistance markers, viruses used as promoters, and genes for producing toxins. Regulating Bt corn for safety needs different systems than regulating organic corn, just as factory farming needs different regulatory processes than free-range chicken.

Diverse production processes and products need laws and science appropriate to them. Chemical processing needs chemistry labs and chemists, GMOs need genetic ID laws, organic processing needs indigenous science and community control. The response of government to the mustard oil contamination in 1998 was to demand that every *ghani* (oil mill) have a lab, a chemist, and must package its oil. This response was inappropriate for the scale and method of production. One million ghanis were shut down, 20,000 small and tiny oil processors were criminalized by an inappropriate law that opened the flood gates for import of soy oil. We cannot allow the destruction unleashed by pseudo safety laws in the edible oils sector to be repeated in other sectors of our indigenous food economy and food culture.

We cannot replace safe systems with unsafe systems through manipulated laws and rules that serve agribusiness, and leave them free to spread food hazards and disease, destroy our diverse foods, and substitute them with unhealthy, hazardous industrial foods. We do not need to deregulate global trade and overregulate domestic production. We need to regulate chemicals and GMOs through centralized structures and regulate local, domestic food systems through local, democratic, decentralized, participatory processes.

The Indian government's proposed unified food regulation law says the Food Authority will take into account international standards. However, in the case of GMOs, there are no international standards, there are the European laws on novel foods and the absolute deregulation of the US. India must craft her laws for her conditions. These laws must be appropriate to the level and content they address. A law for all food systems is a law that privileges large-scale industrial and commercial establishments and discriminates and criminalizes the small, the local, the diverse.

Our kitchens and dhabas, our cottage and household industries are being put in the same category as Nestle, Cargill, ConAgra, and other massive super-industrial processors. Domestic and local consumption, including "not for profit" food provisioning is being put in the same category as imports of hazardous GMOs. This is not a science-based contemporary system. It is an obsolete, crude, and coercive system proposed by a corporate state to destroy 99 percent of our indigenous food processing—and destroy with it millions of livelihoods and millennia of diverse gastronomic traditions—in order that global agribusiness is able to control our entire food economy.

The Right to Information

Coca-Cola and Pepsi are hiding behind trade secrets legislation to block the disclosure of the ingredients of their soft drinks to the Rajasthan High Court. Union Carbide hid behind trade secrets when it refused to disclose the nature of the gas leak in Bhopal and allowed thousands to die and millions to be crippled. Food and health are too important to be sacrificed to corporate confidentiality. The right to information must be the basis of any food safety law.

We need to learn from the food mistakes of the industrialized food systems. These systems have created mad cow disease and unleashed an epidemic of obesity and diabetes. These systems are not identified as "food hazards" in food safety laws, though they are hazardous to health. A modern food law would recognize that our decentralized food economy enhances nutrition, safety, culture, and livelihoods. We need laws to protect our diverse local food cultures from the disease-causing homogenous, centralized, industrial food culture of the West. Our biodiversity and cultural diversity of food are robust localized food economies; our skilled and knowledgeable food processes are the future of food. We cannot allow a law manipulated by global food giants, promoted by power-hungry bureaucrats to take away our food freedom and food sovereignty. We do not need pseudo safety standards that serve global business and destroy our rich food culture. We need society-led, participatory, democratic systems to enrich our food systems, promote health and nutrition, and guarantee food safety. Delhi needs to control the Monsantos and Cargills, not our dhabas and our kitchens. Let the government regulate agribusiness. We will regulate ourselves as a community and civil society. We will not be ruled through the law for food fascism. We will shape laws for our food freedom. This is our food sovereignty. This is our anna swaraj.

Terra Madre: A Celebration of Living Economies

In a world dominated by fear and fragmentation, dispensability and despair, a magical gathering of food communities—Terra Madre—took place in Turin, Italy, in October 2004. Slow Food, the movement that has put the culture of growing and eating good, healthy, diverse food at the heart of social, political, and economic transformation, brought together 5,000 members from 1,200 food communities in 130 countries. Despite the diversity and differences, everyone was connected: connected through the earth, our mother, Terra Madre; connected through food, the very web of life; connected through our common humanity, which makes a peasant the equal of a prince.

Terra Madre was a gathering of small producers who refuse to disappear in a world where globalization has written off diversity of species and cultures, small producers, local economies, and indigenous knowledge. Not only are small farmers and local food communities refusing to go away, they are determined to shape a future beyond globalization. As Giovanni Alemanno, the Italian minister of agriculture and forestry said in his introductory speech at Terra Madre:

> What is original and truly revolutionary about Terra Madre is that by selecting the food communities least susceptible to industrial process—hence distinctive for the authenticity and quality of their produce—it attempts to place small-scale food producers at center stage.

Over the past few decades, food production, processing, and distribution has shifted out of the hands of women, small farmers, and small producers and is being monopolized by global corporate giants such as Cargill, Monsanto, Phillip Morris, and Nestle. Small producers everywhere are being displaced and uprooted by the unfair competition from heavily subsidized agribusiness. The antiglobalization movement has focused on the unfairness of global trade rules that are pushing farmers into debt and suicide. At Terra Madre, small producers gathered not just to curse the darkness of corporate globalization, but also to light and keep lit the lamps of small decentralized, biodiverse production.

The vibrant energy of Terra Madre came from the resilience of producers who had continued to save and share their diverse seeds, live their diverse cultures, speak their diverse languages, and celebrate their diverse food traditions. There was a community of dried mango producers; entomophagous women of Ouagadougou, Burkina Faso (women who harvest, process, and sell edible insects); the Baobab community of Atacora, Benin; basil growers; makers of Liguria; nomadic shepherds from India and Kirgbity; sheep breeders from Central Asia; jasmine rice producers from Thailand; and basmati rice producers from India (the last of these had both been biopiracy victims of RiceTec). The world of Terra Madre reflected the real world of people with diversity so dazzling that the eyes and ears were having a feast, while communities communicated with pride, joy, and dignity about their agricultural and food traditions. This was not the world of the WTO where only agribusiness exists, where agricultural trade basically means soya, corn, rice, wheat, and canola; where one company (Monsanto) accounts for 94 percent of the world's GMO; and where most food grown is not eaten by humans but by billions of captive animals in factory farms. In Terra Madre's world, small farms produce more than industrial farms while using fewer resources; biodiversity protects the health of the soil and the health of people; and quality, taste, and nutrition are the criteria for production and processing, not toxic quantity and superprofits of agribusiness.

Diversity provides the ground for us to turn our food systems around—diversity of crops, diversity of foods, and diversity of cultures. Diversity is both the resistance to monocultures and the creative alternative. Building on our uniqueness and variety is our strength, a strength that can be eroded only when we give up on it ourselves.

Another Paradigm for Food

Terra Madre provided an opportunity and platform to articulate another paradigm for food. During the opening ceremony, Carlo Petrini, the founder of Slow Food, called upon everyone to defend the rights, knowledge, and creativity of small producers all over the world. He also called for us to abandon the gap between consumer and producer. "Let us become coproducers," he said. To consume means to destroy. That's why "consumption" was the name given to tuberculosis. In the act of eating, we are already participating in production. By eating organic, we have said no to toxics and supported the organic farmer. By rejecting GMOs, we vote for the rights of small farmers and people's right to information and health. By eating local, we have taken power and profits away from global agribusiness, and strengthened our local food community. Eaters are, therefore, also coproducers, both because their relationship with small producers is a critical link in creating a sustainable, just, healthy food system, and because we are what we eat, and in making food choices, we make choices about who we are.

The industrialization and globalization of our food systems is dividing us: North-South, producer-consumer, rich-poor. The most significant source of our separation and division is. the myth of "cheap" food, the myth that industrial food systems produce more food and hence are necessary to end poverty. However, small biodiverse, organic farms have higher output than large industrial monocultures.

As Prince Charles of Great Britain reminded the gathering during the closing ceremony:

> One of the arguments used by the "agricultural industrialists" is that it is only through intensification that we will be able to feed an expanded world population. But even without significant investment, and often in the face of official disapproval, improved organic practices have increased yields and outputs dramatically. A recent UN-FAO study revealed that in Bolivia potato yield went up from four to fifteen tons per hectare. In Cuba, the vegetable yields of organic urban gardens almost doubled. In Ethiopia, which twenty years ago suffered appalling famine, sweet potato yields went up from six to thirty tons per hectare. In Kenya, maize yields increased from two and a quarter to nine tons per hectare. And in Pakistan, mango yields have gone up from seven and half to twenty-two tons per hectare.

During the inaugural ceremony, I drew attention to the fact that globalized, industrialized food is not cheap. It is too costly for the earth, for the farmers, and for our health. The earth can no longer carry the burden of groundwater mining, pesticide pollution, disappearance of species, and destabilization of the climate. Farmers can no longer carry the burden of debt inevitable in industrial farming.

The 30,000 farmer suicides in India over a decade are a symptom of the deep crisis in the dominant model of farming and food production. This system is denying the right of food and health both to the 1 billion who are hungry and the 1 billion who suffer from obesity. It is incapable of producing safe, culturally appropriate, tasty, quality food. And it is incapable of producing enough food for all because it is wasteful of land, water, and energy. Corporate agriculture uses 10 times more energy than it produces, 10 times more water than ecological agriculture. It is thus 10 times less efficient. Labor efficiency is also a myth; all the researchers, pesticide producers, genetic engineers, truck drivers, and soldiers engaged in wars over oil are part of the industrial food production system. If all the people involved in nonsustainable food production were counted, the labor efficiency of industrial food world also be lower than ecological food. When agriculture becomes like war, and weapons of mass destruction replace internal farm inputs, food becomes non-food. Trade based on false prices and unfair exchange is not trade, it is exploitation.

Industrial food is cheap not because it is efficient—either in terms of resource or energy efficiency—but because it is supported by subsidies and it externalizes all costs—the wars, the diseases, the environmental destruction, the cultural decay, the social disintegration.

Terra Madre was a celebration of an honest agriculture in which prices do not lie, which does not exploit the earth and the earth's caretakers. Terra Madre was a celebration of our practice of living economies in which we coproduce with the earthworm and the spider, with the mycorrhiza and the fungi. We are all connected in the web of life, and it is food that spins that web. As the ancient *Taitreya Upanishad* tells us: "From food, all creatures are produced. . . . Beings are born from food, when born they live by food, on being deceased, they enter into food."

In India, we are creating food democracy through freedom farms, freedom villages, and freedom zones. Organic farms free of chemicals and toxins and zones free of corporate—that is GMO—and patented seeds are creating a bottom-up democracy of food to counter the top-down food dictatorship.

Food Democracy

The principles for food democracy and food freedom are shared by people across diverse cultures as is evident in the Manifesto of the Commission on the Future of Food which I co-chair with the president of the region of Tuscany in Italy. As the foreword states:

> The manifesto is intended as a synthesis of the work and the ideas espoused by hundreds of organizations around the world and thousands of individuals actively seeking to reverse the present dire trend toward the industrialization and globalization of food production.
> While the manifesto includes a critique of the dangerous directions of the moment, most importantly it sets out practical vision, ideas and programs

> toward ensuring that food and agriculture become more socially and ecologically sustainable, more accessible, and toward putting food quality, food safety and public health above corporate profits.
> We hope this manifesto will serve as a catalyst to unify and strengthen the movement toward sustainable agriculture, food sovereignty, biodiversity, and agricultural diversity, and that it will help thereby to alleviate hunger and poverty globally.[4]

The manifesto details the challenges and goals of what our future food society should look like and then goes on to highlight some "living alternatives to industrial agriculture":

> On every continent, communities are awakening to the devastating effects of corporate-driven food and farming systems that have turned agriculture into an extractive industry and food into a major health hazard. Movements are emerging, many with parallels and linkages across international borders, that are re-knitting the historic relationships among food, farming, and community values. These movements are restoring food and food production to their proper places in culture and nature after a devastating estrangement that stands as an aberration in the human experience.
> Here we only have sufficient space to hint at the breakthroughs these movements have made in the last several decades. The fact that few of these changes could have been predicted in advance should give pause to anyone who now argues that industrial agriculture is the inevitable way forward. Change—very rapid change—is possible. Indeed, it is underway. The following are a few of the areas where circumstances are rapidly changing:
>
> *Democratizing access to land*
>
> While it has long been recognized that access to land by the world's rural poor is a key to ending hunger and poverty, many believed reform to be politically impossible. This was true in Brazil, where less than two percent of rural landholders held half the farmland (most of it left idle), and where even small gatherings were outlawed and efforts for change were met with violence. Yet today this country leads the way toward democratizing access to land. During the last 20 years, the Landless Workers' Movement, called by its Portuguese acronym MST, has settled a quarter-million formerly landless families on 15 million acres of land in almost every state of Brazil. Taking advantage of a clause in the new constitution mandating the government to redistribute unused land, the MST has used disciplined civil disobedience to ensure this mandate's fulfillment.

The MST's almost 3,000 new communities are creating thousands of new businesses and schools. Land reform benefits are measured in an annual income for new MST settlers of almost four times the minimum wage, while still-landless workers now receive on average only 70 percent of the minimum. Infant mortality among land reform families has fallen to only half the national average. Estimates of the cost of creating a job in the commercial sector of Brazil range from two to 20 times more than the cost of establishing the unemployed family on the land through land reform. Democratizing access to land is working.

Democratizing access to credit

Bankers long held that poor people were unacceptable credit risks. But that barrier is falling. In Bangladesh two decades ago, the Grameen Bank created a rural credit system based not on property collateral but on small-group mutual responsibility. Grameen's microcredit loan program, made to 2.5 million poor villagers, mostly women, has been adopted in 58 countries. With a repayment rate far superior to traditional banks, democratizing access to investment resources is proving viable.

Re-linking city and country, consumer and grower

On every continent, practical steps are underway to make local production for local consumption viable. "Buy local" campaigns are appealing to consumers in Europe, the US, and elsewhere. One innovation is the Community-Supported Agriculture (CSA) movement in which farmers and consumers link and share risks. Consumers buy a "share" at the beginning of the season, entitling them to the fruits of the farmers' labors. CSAs emerged in the mid-60s in Germany, Switzerland and in Japan. Seventeen years ago, no CSAs existed in the US; today, there are more than 3,000 serving tens of thousands of families. The US example has helped inspire a CSA movement in the United Kingdom, which has won local government support. Similar movements have simultaneously developed in Japan and elsewhere.

Other burgeoning initiatives are urban and rural farmers' markets, which have grown by 79 percent in the last eight years in the US alone. These have enabled local farmers to sell directly to their publics without expensive intermediaries. Small gardens from family kitchen gardens in Kenya to school gardens enabling children to grow their own meals in California are also spreading.[5]

Jal Swaraj [Water Democracy]

Water is life. Without water democracy there can be no living democracy.

The biosphere is a biosphere because it is a hydrosphere. The planet's hydrological cycle is a water democracy—a system of distributing water for all species—the rain forest in the Amazon, the desert life in the Sahara. Nature does not distribute water *uniformly*. It distributes it *equitably*. Uniformity would mean each part of the planet has the same amount of precipitation, in the same quantity, and the same pattern. It would mean the same plants grow across the planet, the same species are found everywhere. But the planet creates and maintains diversity, and this diversity evolves because of diversity in water regimes. However, within each ecosystem, each agro-climatic zone, water is equitably distributed—all species get their share of water. Nature does not discriminate between the needs of a microbe and a mammal, plants and humans. And all humans as a species have the same sustenance needs for water.

Globalization is undermining the planet's water democracy through overexploitation of groundwater, rerouting and diverting of rivers, and privatization of public supply. Since I wrote *Water Wars*, the projects, policies, and processes of water privatization and commodification are much more evident, and the movements for water democracy are much more pervasive. I wrote about CocaCola and Pepsi and their water grab. Across India, movements have emerged to resist their water theft and toxic sales. I wrote about dams. There are new plans to reroute all of India's rivers in a major river-linking project and movements are growing to fight these plans. I wrote about World Bank–driven privatization. We are in the midst of resistance to the privatization of Delhi's water supply. The stories about the struggles for water democracy in India are stories being repeated everywhere in the world. Whether it is water mining by the cola giants, river diversion projects, or privatization of urban supply, corporate hijacking of water is facilitated by the creation of corporate states—states that centralize power, destroy federal structures and the constitutional fabric, and usurp and erode fundamental community rights.

Struggles of water democracy against corporate giants thus also became struggles against centralizing states. Without centralized state control, privatization is not possible. The market rules through coercive, anti-people, undemocratic states. That is why Earth Democracy, and one of its facets, water democracy, is simultaneously a deepening of democracy and a defense of genuinely democratic structures. It is simultaneously a process of reclaiming the commons and community rights and defending common public goods and public services.

Women Against Coca-Cola

Women in a small hamlet in Kerala succeeded in shutting down a Coca-Cola plant. "When you drink Coke, you drink the blood of people," said Mylamma, the woman who started the movement against Coca-Cola in Plachimada.[6]

The Coca-Cola plant in Plachimada was commissioned in March 2000 to produce 1,224,000 bottles of Coca-Cola products a day and issued a conditional license to install a motor-driven water pump by the panchayat. However, the company started to illegally extract millions of liters of clean water. According to the local people, Coca-Cola was extracting 1.5 million liters per day. The water level started to fall, dropping from 150 to 500 feet below the earth's surface. Tribals and farmers complained that water storage and supply were being adversely affected by indiscriminate installation of bore wells for tapping groundwater, resulting in serious consequences for crop cultivation. The wells were also threatening traditional drinking-water sources, ponds and water tanks, water ways and canals. When the company failed to comply with the panchayat request for details, a show-cause notice was served and the license was cancelled. Coca-Cola unsuccessfully tried to bribe the panchayat president A. Krishnan, with 300 million rupees.

Not only did Coca-Cola steal the water of the local community, it also polluted what it didn't take. The company deposited waste material outside the plant which, during the rainy season, spread into paddy fields, canals, and wells, causing serious health hazards. As a result of this dumping, 260 bore wells provided by public authorities for drinking water and agriculture facilities have become dry. Coca-Cola was also pumping wastewater into dry bore wells within the company premises. In 2003, the district medical officer informed the people of Plachimada their water was unfit for drinking. The women, who already knew their water was toxic, had to walk miles to get water. Coca-Cola had created water scarcity in a water-abundant region.

The women of Plachimada were not going to allow this hydropiracy. In 2002 they started a *dharna* (sit-in) at the gates of Coca-Cola. To celebrate one year of their agitation, I joined them on Earth Day 2003. On September 21, 2003, a huge rally delivered an ultimatum to Coca-Cola. And in January 2004, a World Water Conference brought global activists like Jose Bove and Maude Barlow to Plachimada to support the local activists. A movement started by local adivasi women had unleashed a national and global wave of people's energy in their support.

The local panchayat used its constitutional rights to serve notice to Coca-Cola. The Perumatty panchayat also filed publicinterest litigation in the Kerala High Court against Coca-Cola. The court supported the women's demands and, in an order given on December 16, 2003, Justice Balakrishnana Nair ordered CocaCola to stop pirating Plachimada's water. Justice Nair's decision stated:

> The public trust doctrine primarily rests on the principle that certain resources like air, sea, waters, and the forests have such a great importance to the people as a whole that it would be wholly unjustified to make them a subject of private ownership. The said resources being a gift of nature, they should be made freely available to everyone irrespective of their status in life. The doctrine enjoins upon the government to protect the resources for the enjoyment of the general public rather than to permit their use for private ownership or commercial purpose....
>
> Our legal system—based on English common law—includes the public trust doctrine as part of its jurisprudence. The State is the trustee of all natural resources, which are by nature meant for public use and enjoyment. Public at large is the beneficiary of the seashore, running waters, airs, forests, and ecologically fragile lands. The State as a trustee is under a legal duty to protect the natural resources. These resources meant for public use cannot be converted into private ownership.[7]

On February 17, 2004, the Kerala chief minister, under pressure from the growing movement and a drought-aggravated water crisis, ordered the closure of the Coca-Cola plant. The victory of the movement in Plachimada was the result of creating broad alliances and using multiple strategies. The local movement of women in Plachimada triggered recognition of people's community rights to water in law, while also triggering movements against the 87 other Coca-Cola and Pepsi plants where water is being depleted and polluted.

Plachimada Declaration

Water is the basis of life; it is the gift of nature; it belongs to all living beings on earth.
Water is not private property. It is a common resource for the sustenance of all.
Water is the fundamental human right. It has to be conserved, protected, and managed. It is our fundamental obligation to prevent water scarcity and pollution and to preserve it for generations.
Water is not a commodity. We should resist all criminal attempts to marketize, privatize, and corporatize water. Only through these means we can ensure the fundamental and inalienable right to water for people all over the world.
The Water Policy should be formulated on the basis of this outlook.
The right to conserve, use, and manage water is fully vested with the local community. This is the very basis of water democracy. Any attempt to reduce or deny this right is a crime.
The production and marketing of the poisonous products of the Coca-Cola and Pepsi Cola corporations lead to total destruction and pollution and also endangers the very existence of local communities.
The resistance that has come up in Plachimada, Puduchery, and in various parts of the world is the symbol of our valiant struggle against the devilish corporate gangs who pirate our water.
We, who are in the battlefield in full solidarity with the adivasis who have put up resistance against the tortures of the horrid commercial forces in Plachimada, exhort the people all over the world to boycott the products of Coca-Cola and Pepsi Cola.

Coca-Cola–Pepsi Cola "Quit India"

Plachimada created new energy for local resistance everywhere. In May 2004, groups from across India fighting against water mining met in Delhi to coordinate their actions as the Coca-Cola–Pepsi Quit India Campaign.

Every plant is a Plachimada in the making. Coca-Cola set up a plant in Kaladere, Jaipur, the capital of Rajasthan, in 1999. The water table has since dropped from 40 to 125 feet, leaving wells and hand pumps dry. The protest against Coca-Cola has been growing there, and an eminent Gandhian, Siddharaj Dodda, was arrested when he joined a peaceful march to demand the closure of the plant. In its ruling on a public-interest case, the Rajasthan High Court ordered the cola giants to stop sales for refusing to disclose the contents of their products. Coca-Cola and Pepsi challenged the High Court's order in the Supreme Court. The Supreme Court ruled against the soft drink giants and ordered disclosure according to the Rajasthan ruling. The Center for Public Interest Litigation, through its lawyer Prashant Bhushan, has filed a writ demanding full disclosure on the basis of the toxic hazards contained in the drinks.

On December 2, 2004, the 20th anniversary of the Bhopal tragedy, a major conference called "Detoxification" drew connections between the toxic leak at the Union Carbide pesticide plant in 1984; the continued spread of toxins in agriculture through pesticides, herbicides, and GMOs; and the toxins in soft drinks sold by Coca-Cola and Pepsi, which farmers have shown to be effective as pesticides.

In Mehdiganj, 20 kilometers from the holy city of Varanasi, villagers are protesting against the Coca-Cola plant. The water table has dropped by 40 feet, and the fields around the plant are polluted. On May 10, 2003, a hundred people demonstrated, and Jagrupa Devi, an elderly woman, was sent to the hospital bleeding with head injuries. On September 10, 2003, 500 people protested; 14 were injured, were arrested. In October 2003, a case was filed against Coca-Cola for stealing people's water. On January 20, 2005 thousands of people blockaded Coca-Cola and Pepsi plants across India and served notices to the cola giants to stop water theft. Hundreds of schools and colleges have declared themselves "Coke-Pepsi free zones."

In the building of Earth Democracy, the first step is to take action for one's own freedom. Rights flow from taking responsibility. Freedom grows by living free.

Creating Water Democracy in Delhi

Delhi, India's capital, has been sustained for centuries by the river Yamuna. The 16th-century poet Sant Vallabhacharya wrote the *Yamunastakam* in praise of the Yamuna.

> I bow joyfully to Yamuna, the source of all spiritual abilities.
> You are richly endowed with innumerable sands glistening from contact with lotus-feet of Krishna.
> Your water is delightfully scented with fragrant flowers from the fresh flowers from the fresh forests that flourish on your banks.

You bear the beauty of Krishna, Cupid's father, who is worshipped by both the gods and demons.

You rush down from Kalinda Mountain, your waters bright with white foam.
Anxious for love you gush onward, rising and falling through the boulders.
Your excited, undulating motions create melodious songs, and it appears that you are mounted on a swaying palanquin of love.
Glory be to Yamuna, daughter of the sun, who increases love for Krishna.

You have descended to purify the earth.
Parrots, peacocks, swans, and other birds serve you with their various sons, as if they were your dear friends.
Your waves appear as braceleted arms, and your banks as beautiful hips decorated with sands that look like pearl-studded ornaments.
I bow to you, fourth beloved of Krishna.
You are adorned with countless qualities, and are praised by
Siva, Brahma, and other gods.[8]

Two decades of industrialization have turned the Yamuna into a toxic sewer. Instead of stopping the pollution, the World Bank, using the scarcity created by the pollution, pushed the Delhi government to privatize Delhi's water supply and get water from the Tehri Dam on the Ganges, hundreds of miles away. A privatized plant that could have been built for 1 billion rupees has cost the public 7 billion rupees.

The privatization of Delhi's water supply is centered around the Sonia Vihar water treatment plant. The plant, which was inaugurated on June 21, 2002, is designed at a cost of 1.8 billion rupees for a capacity of 635 million liters a day on a 10-year build-operate-transfer (BOT) basis. The contract between Delhi Jal Board and the French company Ondeo Degremont (a subsidiary of Suez Lyonnaise des Eaux Water Division—the water giant of the world), is supposed to provide safe drinking water for the city.

The water for the Suez-Degremont plant in Delhi will come from the Tehri Dam through the Upper Ganga Canal to Muradnagar in western Uttar Pradesh and then through the giant pipeline to Delhi. The Upper Ganga Canal, which starts at Haridwar and carries the holy water of the Ganga to Kanpur via Muradnagar, is the main source of irrigation for this region.

Suez is not bringing in private foreign investment. It is appropriating public investment. Public-private partnerships are, in effect, private appropriation of public investment. But the financial costs are not the highest costs. The real costs are social and ecological.

The Ganga is also being transformed from a river of life to a river of death by the ecological consequences of damming and diversion. The Tehri Dam, located in the outer Himalaya, in the Tehri-Garhwal district of Uttaranchal, is planned to be the fifth highest dam in the world. If completed, it will be 260.5 meters high and create a lake spread over an area of 45 square kilometers of land in the Bhagirathi and Bhilangana valleys. The dam will submerge 4,200 hectares of the most fertile flat land in those valleys without benefiting the region in any way.

Additionally, the area is earthquake prone and the huge Tehri Dam is located in a seismic fault zone. Between 1816 and 1991, there have been 17 earthquakes in the Garhwal region, with recent ones occurring in Uttarkashi in 1991 and Chamoli in1998. The International Commission on Large Dams has declared the dam site "extremely hazardous."

If the dam collapses from an earthquake—or from any other fault, such as a landslide—the devastation will be unimaginable. The huge reservoir will be emptied in 22 minutes. Within an hour Rishikesh will be under 260 meters of water. Within the next 23 minutes Haridwar will be submerged under 232 meters of water. Bijnor, Meerut, Hapur, and Bulandshahar will be under water within 12 hours.[9] The dam is potentially dangerous for large parts of northwestern India, and large areas in the Gangetic Plain could be devastated.[10]

Already, the islands of silt are rising faster than the captured water. It is estimated that the life of the dam would not be more than 30 years because of the heavy sedimentation. The Tehri Dam will hold silt, not water, and create floods, not prevent them.

Diversion too spells catastrophe. The disappearance of the Ganga in the peak of the summer of 2003 was an experiment—a vivisection of our living rivers, our living cultures—allegedly to clean the "ghats" at Haridwar, but designed to test how much violence as a society we will tolerate as mute witnesses to our own destruction. The people of Uttaranchal, Uttar Pradesh, and Delhi can turn around this violent, abusive experiment and transform the conversion of the lifeblood of our rivers from corporate commodities into an experience for ensuring water justice and sustainability.

The people of Tehri can never be compensated for the uprooting of their lives. The women are still sitting on a *dharna*, refusing to move, even though the government paid contractors to break down the homes to force the people to move. All local water development projects in the darn catchment area have been canceled on the grounds that the government has no money and because every drop of Ganga water must flow into the dam. Nearly a hundred women are said to have committed suicide in the Pratap Nagar area for lack of water, even though the Ganga flows below their villages. As one woman declared, "The Ganga, which was our mother, has become our graveyard." Privatization of water denies local communities their water rights and access to water.

On December 1, 2004, water tariffs were increased in Delhi. While the government stated the increase was necessary for recovering costs of operation and maintenance, the tariff increase is ten times more than what is needed to run Delhi's water supply. The reason for the increase is to lay the ground for the privatization of Delhi's water and ensure superprofits for the private operators. Increasing tariffs *before* privatization is part of the World Bank's tool kit. In preparation for full privatization, the "private-public partnership" increases tariffs of public utilities, in order to "support a commercial operation," that is, guarantee profit margins. Service and management contracts can be introduced while the government increases tariffs. The tariff increase is not a democratic or a need-based decision. It has been imposed by the World Bank. The Delhi Jal Board (the city's water board) cites a study on privatization done by Price Waterhouse Cooper under the auspices of the World Bank as the justification for increase in tariff in addition to a World Bank technical paper on water pricing.[11]

Delhi's water operation and maintenance budget is 3.44 billion rupees. The public utility has recovered 2.7 billion rupees due to nonrevenue losses such as leaks and thefts. A conference on "public-public partnership" showed how public and community participation can recover an additional 5 billion rupees by preventing leaks and theft. By preventing leaks and theft through people's participation, the public water utility can increase water availability as well as its income. This 7 to 8 billion rupee recovery is twice the amount needed to operate and maintain the water system.

However, the tariff increase will allow a recovery of 30 billion rupees, tenfold more than needed, guaranteeing a profit of more than 26.66 billion rupees to the corporations waiting to grab Delhi's water supply. An extra 10 percent increase is built into the tariff restructuring, which will double the profits for water privateers in seven years. This profit is created not by providing better services, but by doubling the financial burden on citizens, especially the poor.

Changes in categories hide significant tariff increases. Schools and agriculture have been redefined as industry. *Piaos*, public, free water stalls which form a core part of India's gift of water culture, must also pay for water. How will they give water to the thirsty? Cremation grounds, temples, homes for the disabled, and orphanages which previously paid 30 rupees will now have to pay thousands of rupees. The cash-strapped social institutions cannot pay. The World Bank-driven policies explicitly state that there needs to be a shift from social to commercial value. This worldview lies at the root of the conflict between water privatization and water democracy.

Many privatization myths have been used to justify the tariff increases. The first is the myth of full cost recovery. The privaters logic is that water tariffs need to be increased because the full cost of providing water must be recovered. However, as far as operations are concerned, the tariff increase implies a "tenfold recovery," that is, 10 times more than full cost. However, as we have shown, private companies do not make investments. Investments are made by the public, both in terms of setting up water systems historically and in terms of the World Bank loans that impose privatization. The private operators have made no investment, but will harvest 1 trillion rupees of public investment. Therefore, when full costs are accounted for, the public has made the investments and water systems must stay in public hands as a common good.

The second myth of water privatization is that it will improve services to the poor. Delhi's new supply from Sonia Vihar will go to the rich South Delhi areas, not to the slums. The claim that the poor will get 40 liters per day for free as a "lifeline" is also false, since the new tariff structure will collect 40 rupees as a flat rate even from the slums where homes do not have water connections.

A third myth is that increased tariffs will lead to reduced use. This myth is a corollary to the myth that water waste is linked to water not being priced on the market. Women who walk 10 miles for water do not waste a drop, even though their water is not provided through market transactions. And the rich can waste water in spite of increased tariffs because compared with their incomes, the increase is insignificant. Privatization rewards this waste. The project to supply water to rich colonies 24/7 is an encouragement to waste. This waste does not just divert water from the poor in Delhi, it diverts water from other regions.

Delhi's ever growing water demands have already led to major diversions of water from other regions. Delhi already gets 455 million liters from the Ganga. With the Sonia Vihar plant's demand for 635 million liters, 1,090 million liters per day are diverted from the Ganga. Further diversions of three billion cubic meters from the Ganga are built into the Sharda and Yamuna river link. Delhi is also demanding 180 million liters per day to be diverted from Punjab's Dhakra dam. Water will also be diverted to Delhi from the Renuka darn on the Giri river (1.25 billion cubic liters per day) and Keshau dam on the Tons river (610 million cubic liters per day).

These diversions will have huge ecological and social costs. On June 13, 2005, five farmers were shot protesting the diversion of water from Bisalpur dam for Jaipur city through an Asian Development Bank project. The mega diversion for water by the rich in Delhi could trigger major water wars.

Building water democracy means building alliances. When advertisement for the inauguration of Suez's Sonia Vihar plant appeared on June 2, 2002, I started to contact citizens groups in Delhi and people's movements along the Ganges. Each group helped frame the struggle against privatization and everyone's issue became a key to resistance. The 100,000 people displaced by Tehri Dam were linked to the millions of Indians who hold the Ganges as sacred, who, in turn, were connected to farmers whose land and water would be appropriated. Millions signed petitions saying, "Our Mother Ganga is not for sale." We organized a Jal Swaraj Yarra (a water democracy journey) from March 15 to 22, World Water Day. We did Ganga Yatras to rejuvenate the living culture of the sacred Ganges. A million people were reached; 150,000 signed a hundred-meter "river" of cloth to protest privatization. The government of Uttaranchal (where the Tehri Dam is located) and the government of Uttar Pradesh (from where the water was to be diverted) refused to supply water to the Suez plant in Delhi.

We do not need privatization or river diversions to address Delhi's water problems. We have shown how with equitable distribution and a combination of conservation, recycling, and reduction in use, Delhi's water needs can be. met locally. We need democracy and conservation. The seeds for the water democracy movement in Delhi have been sown. We now have to nurture them to reclaim water as a commons and a public good.

Citizens' Statement on Water Tariff Increase

The citizens of Delhi are committed to conservation and equitable use of our scarce, but precious water resources.

We are also committed to defend our fundamental right to water, which can only be protected through a public system which treats water as a public good and essential service. Our right to water propels our pledge to keep Delhi's water supply in the public domain.

We condemn the anti-democratic, unjustified hike in tariffs announced by the Delhi government on November 30, 2004, which is a preparation for water privatization.

Citizens have offered models for public-private partnership to reduce waste and reduce costs, and provide safe, clean, affordable water for all.

The tariff increase carried out by the Delhi Jal Board on World Bank dictate will promote waste by the rich and put a burden on the poor. This "rationalization" might suit World Bank rationality to privatize and commodify the last drop of water. However, it goes against our culture and our constitutional rights.

The Government and World Bank are paving the way for Multinational Corporations (MNC) like Suez to take control over our water. The contract for the Sonia Vihar plant has already been given to Suez-Degremont. With the tariff increase, a profitable "water market" is being created for MNCs.

As a brand new citizens' alliance of residents, environmental groups, religious groups, health groups, water workers, we will continue to work creatively and constructively to defend "water for life, not for profits." We will not allow our water to be hijacked. We will not let our democratic rights be bypassed. We will not let our fundamental right to water be eroded.

Toward that goal we created the Citizens Front for Water Democracy on December 1, 2004, in response to the announcement of the tariff increase.

Rerouting Rivers: The Dream Project for Water Privateers

Free-flowing rivers are free, in the sense that they do not need capital investment, they are not enclosed, and their waters are accessible to all. Water locked in dams and canals are captive waters. They can be privatized, commoditized, bought, sold, and controlled by the powerful.

The massive $200 billion River Linking Project, supported in part by the World Bank, is a key to the privatization of water and the enclosure of India's water commons.

The River Linking Project is divided into two broad components: the Himalayan component and the Peninsular component. The Himalayan part consists of 14 river links with an estimated cost of 3.75 trillion rupees, and the Peninsular component consists of 16 river links, estimated at 1.85 trillion rupees.

The Himalayan Component

Kosi–Mechi Link
Kosi–Ghaghra Link
Gandak–Ganga Link
Ghaghra–Yamuna Link
Sarda–Yamuna Link
Yamuna–Rajasthan Link
Rajasthan–Sabarmati Link
Chunar–Sone Barrage Link
Sone Dam–Southern Tributaries of Ganga Link
Brahmaputra–Ganga Link (Manas–Sankosh Tista–Ganga)
Brahmaputra–Ganga Link Jogighopa Tista–Farakka)
Farakka–Sunderbans Link
Ganga–Damodar–Sundernarekha Link
Subernarekha–Mahanadi Link

The Peninsular Component

Mahanadi (Manibhadra)–Godavari (Dowlaiswaram) Link
Giodavari (Polavarm)–Krishna (Vijayawada) Link
Godavari (Inchamaplli)–Krishna (Nagarjunasagar) Link
Godavari (Inchampalli Low Dam)–Krishna
(Nagarjunasagar Tail Pond) Link
Krishna (Nagarjunasagar)–Pennar (Somasila) Link
Krishna (Srisailam)–Pennar Link
Krishna (Almatti)–Pennar Link
Pennar (Somasila)–Cauvery (Grand Anicut) Link

Cauvery (Kattalai)–Vaigai (Gundar) Link
Parbati—Kalishindh–Chambal Link
Damanganga–Pinjal Link
Par–Tapi–Narmada Link
Ken–Betwa Link
Pamba–Achankovil–Vaippar Link
Netrreavati–Hemavati Link
Bedti–Varda Link

Even the minimum estimated cost of 5.6 trillion rupees equals one-quarter of our annual gross domestic product (GDP), two-and-a-half times our annual tax collection, and twice our present foreign exchange reserves. The cost of this project, according to the government's economic survey for 2001–2002, is higher than India's gross domestic savings and more than $12 billion higher than India's total outstanding external debt.[12] Where is capital of this magnitude available?

The only option would be funds from international sources. Such funding would place a debt of about $112 on every Indian—20 percent of the average annual income. The annual interest would range between 200 and 300 billion rupees. It also raises questions about how this loan would be repaid and what guarantees will be needed to secure it.[13]

External borrowing on this scale would also make future governments more vulnerable to foreign financial pressures.

The real threat is that after starting the project with much fanfare and investing thousands of crores into it, a future government would have to simply abandon the project as its financial implications become clear, leaving billions of cubic meters of earth dug up and the face of the country scarred for centuries. In that scenario the only alternative left would be to hand over the project—along with the country's entire water resources to multinational corporations.

The Research Foundation has studied the first link—the Ken–Betwa link—which is being financed by the World Bank. The Ken–Betwa river linking project includes constructing a 73-meter-high dam on the Ken in Bundelkhand, on the border of the Chhattarpur and Panna districts, and a 231-kilometer-long canal which will connect the Ken and Betwa. Seventy-five percent of the estimated 20 billion rupees will be extracted from the local peasants out of various taxes imposed over the next 25 years. That is why the government is proposing such crops, which are water intensive, leading to a hike in water tax.[14]

Fifty square kilometers of Panna Tiger National Park would be submerged by this interlinking project. This national park, through which the Ken flows, is a natural homeland of 10 endangered species listed under Schedule 1 of the 1972 Wildlife Protection Act. This interlinking and transfer of water will affect not only these animal species but also the vegetation, as hundreds of thousands of trees would be cut.

This project proposes five dams altogether, one on the Ken and four on the Betwa, which would displace around 18 villages. All five dams are proposed to be built in protected forest area. The four dams on the Betwa would submerge 800 hectares of forest.

While the Indian government wants to transfer 1 billion cubic meters of water to the Betwa, the Irrigation Departments of Uttar Pradesh and Madhya Pradesh are of the opinion that the Ken does not have that much water. The government's aim of transferring water from the so-called surplus areas to water-short areas cannot be realized. Moreover, as the Ken does not have surplus water and as the Betwa has enough water, the Ken and Betwa river belts would get affected by drought and flood respectively.

As arithmetic the plan appears to balance—water taken out for storage is returned. However, the plan removes water during drought and returns it during flooding, aggravating both. The study by RFSTE shows that even after escalating drought in 40 villages, affecting 75,000 hectares of land in the district of Banda, and flooding in 200 villages, causing the devastation in 400,000 hectares of land in Hamirpur district, in an attempt to fill the link canal it will remain dry for four months during summer. Unique species of fish, each one known throughout India by the name of the pond or lake in which it lives, will be lost. Five thousand fishermen in Chhattarpur and 15,000 in Tikamgarh, whose livelihood is dependent on these fisheries, will face severe famines.

As a result of the water democracy movement in the Bundlekhand region, the government of Uttar Pradesh has refused to transfer the water of the Ken and the people of the Ken basin are determined to resist the river-linking project. Every village in the basin has passed a resolution to declare that water is a commons and that community rights have to be the basis of any water plan or project. Furthermore, the Ken–Betwa link canal would go through places where traditional irrigation has been available for the last 500 years. One district that would get a new irrigation canal is Tikamgarh, but through the ponds constructed by the kings of Chandel and Bundel dynasties, Tikamgarh is already one of the most irrigated agricultural areas in the whole of Bundelkhand.

We organized a water parliament on July 23, 2003, at Orchha, to launch the non-cooperation campaign against the River Linking Project. The entire region has been charged with the spirit of water democracy. As a local organizer said to me, "They destroyed Iraq with bombs. But patents on seeds and diversion of rivers are also bombs that will destroy us. That is why we must resist them." Far away from the glare of global media, ordinary people are making history, not by organizing arms to fight a brutal empire but by self-organizing their lives—their resources, their cultures, their economies—to defeat the empire by turning their back to it, rejecting its tools and its logic, refusing its chains and its dictatorship.

Freedom is being reborn in our villages, from within the community. The torch of freedom is being carried by people, in peace, in partnership. Military force cannot bring us freedom. It never has.

Conclusion

During a recent debate, David Pearce, a World Bank economist who has argued for the commodification of our fast-disappearing natural wealth as a way of conserving resources, admitted that the ecological crisis is deep, and is deepening. But he went on to support privatization of water, commodification of life, and globalization of agriculture. "Large-scale problems," he said, "need large-scale solutions."[15]

However, as Gandhi showed in his life, and as we experience in Earth Democracy, small-scale responses become necessary in periods of dictatorship and totalitarian rule because large-scale structures and processes are controlled by the dominant power. The small becomes powerful in rebuilding living cultures and living democracies because small victories can be claimed by millions. The large is small in terms of the range of people's alternatives. The small is large where unleashing people's energies are concerned.

Gandhi did not bring the British Empire down with cannons or armies that matched those of the imperial forces. He brought down the empire with a pinch of salt and a spinning wheel. When the British introduced the Salt Laws to tax salt, we undertook the Dandi March, picked up salt, and said, "Nature gives it free. We need it for our survival. We will continue to make our own salt. We will disobey the salt laws." When the

Indian textile industry was being destroyed by the British, Gandhi did not call for the industrialization of India's textiles. He pulled out a spinning wheel. As he said, "Anything that millions can do together, becomes charged with power." The spinning wheel became a symbol of such power.

For us our seeds, our rivers, our daily food are sites for reclaiming our economic, political, and cultural freedoms because these are the very sites of the expanding corporate empire over life. We are fully aware that creating living economies based on self-organization and living democracies based on self-rule demands the commitment and courage to resist and disobey the unjust laws that render self-rule, self-provisioning, and self-sustenance illegal. Farmers are forced into corporate slavery by making seed saving illegal. The poor are forced into water markets by privatization contracts. We are all pushed into food fascism by laws that destroy local production and processing.

If we accept these illegal, illegitimate laws, structures, and rules, we will lose our freedom—our living cultures and democracies. As Gandhi taught, freedom can be reclaimed only by refusing to cooperate with unjust, immoral laws. The fight for truth—employing the principles of civil disobedience, nonviolence, and noncooperation—is not just our right as free citizens of free societies. It is our duty as citizens of the earth.

Corporate globalization and militarism go hand in hand. And both are offered through propaganda and a war on truth as recipes for our security. Similarly, Monsanto must declare a war on truth in order to sell us unnecessary, unreliable, genetically engine red seeds. Coca-Cola must declare a war on truth to steal our water. The US government must declare a war on truth to rob our civil liberties in the name of "homeland security." The World Bank must declare a war on truth to lead poor countries and poor people into debt. With Paul Wolfowitz appointed the new president of the World Bank, the common agenda of economic and imperial warfare is made more evident.

To stay free in a period when slavery is sold through spin and propaganda implies that satyagraha, the struggle for truth, must extend to the instruments that colonize our minds and thoughts.

Earth Democracy offers new freedoms to act, but it also offers new freedoms to think—to think of homeland security in terms of our real home—the earth, and in terms of our real security—the ecological security that the earth provides, and the social security that we create through community, through public systems, through shared wealth.

Earth Democracy shifts the worldview from one dominated by markets and military, monocultures and mechanistic reductionism, to the peaceful cocreation and coevolution of diverse beings, connected through the common bonds of life.

Whether it is the commodification of life through an economics of scarcity or the rule of terror through the politics of insecurity, separation, and disconnection—both are used to dominate. Earth Democracy allows us to overcome artificial scarcity and manipulated and manufactured insecurities by seeing and experiencing connections. We begin to see the connections between corporations and corporate states, the connections between economic wars and military wars, the connections between corporate profits and people's poverty, the connections between economic globalization and religious fundamentalism. We also start to discover the connections we have to the earth and to one another. Exposing the connections of dominant powers enables us to evolve appropriate strategies to transform dead democracies into living democracies. Our ecological and social connectedness enables us to create living economies and living cultures, while building the solidarities that crack open the alliances of the powerful. We are poor, insecure, and not free when we are atomized, trapped, divided, and blind to our multiple potentials as earth citizens. We have the potential to participate creatively in building alternatives to the systems designed for total control and limitless profits.

Earth Democracy allows us to remove our blinders, imagine and create other possibilities. The project of multinational corporate rule does not just extinguish our fundamental freedoms. It threatens to annihilate the very conditions of life for large numbers of humans and other species. Liberation in our genocidal times is, first and foremost, the freedom to stay alive. In this epic contest of the forces of life versus antilife, movements of social and economic justice, ecological sustainability, peace, democracy, and cultural freedoms are all making their diverse but significant contributions. Dictatorship is no longer partial. It engulfs entire economic, political, and cultural lives in every society and every country. Freedom too can no longer be partially fought for and defended. Earth Democracy empowers us to create and defend our indivisible and diverse freedoms, while we engage in our particular struggles with our particular passions. Imperialism has always had global reach. Today's movements have a planetary reach and a planetary embrace. We have just begun to tap our potential for transformation and liberation. This is not the end of history, but another beginning.

NOTES

CHAPTER ONE: Living Economies

1. Mahatma K. Gandhi quoted in Vandana Shiva, *Tomorrow's Biodiversity* (London: Thames & Hudson, 2000), 131.
2. Mathew Fox, *The Reinvention of Work: A New Vision for Livelihood in Our Time* (New York: Harper Collins, 1994), 141.
3. Robert Frost, *Robert Frost's Poems* (New York: St, Martin's, 2002), 163.
4. A *privateer* traditionally refers to a state-sanctioned pirate on the open sea. I use it to refer to the person enclosing a common because they are privatizing a resource and engaging in piracy against the people.
5. G. Elliot Smith quoted in Jeremy Rifkin, *Biosphere Politics: A New Consciousness for a New Century* (New York: Crown, 1991), 39.
6. Thomas More quoted in Rifkin, *Biosphere Politics,* 41.
7. Rifkin, *Biosphere Politics,* 41.
8. Boyle quoted in Anthony McCann, *Beyond the Commons: The Expansion of the Irish Music Rights Organisation, the Elimination of Uncertainty, and the Politics of Enclosure* (PhD diss., University of Limerick, 2002), 236.
9. Neeson quoted in McCann, *Beyond the Commons,* 236.
10. George Sturt quoted in Kirkpatrich Sale, *Rebels Against the Future: The Luddites and Their "War on the Industrial Revolution: Lessons for the Computer Age* (Boston: Addison Wesley, 1995), 35.
11. Baden H Baden-Powell, *Land Systems of British India: Being a Manual of the Land-Tenures and of the Systems of Land-Revenue Administration Prevalent in the Several Provinces* (London: Oxford, 1907).
12. The Famine Inquiry Commission, *The Famine Inquiry Commission Report on Bengal* (Calcutta, 1944, reprinted New Delhi: Usha, 1984), 27.
13. J. N. Uppal, *Bengal Famine of 1943: A Man-Made Tragedy* (Delhi: Lucknow, 1984), 60.

14. John Winthrop quoted in Djelal Kadir, *Columbus and the Ends of the Earth* (Berkeley: University of California Press, 1992), 171.
15. Vandana Shiva, *Biopiracy: The Plunder of Nature and Knowledge* (Boston: South End Press, 1997).
16. Commons have often absorbed people dispossessed by enclosures of diverse kinds. When peasants are dispossessed and displaced by dams, they turn to the forests as commons for survival. Even the urban slum is a de facto commons where the displaced try to survive without formal property rights.
17. Quoted in Rifkin, *Biosphere Politics,* 45.
18. Sale, *Rebels Against the Future,* 35.
19. Vandana Shiva and Radha Holla Bhar, *Sharing the Earth's Harvest: An Ecological History of Food and Farming in India, Vol. 2* (Delhi: RFSTE, 2001), 15.
20. Dharampal, *Despoliation and Defaming of India* (Goa: Other India Press, 1999), 24.
21. Radha Kamal (Radhakarnal) Mukherjee, *Economic History of India* (Allahabad, India: Kitab Mahal, 1967), 183.
22. Ibid., xvii.
23. *Report of the Indian Irrigation Commission, 1901–03* (Calcutta: Government Printing, 1904, reprinted Calcutta: F. K. L. Mukhopadhyay 1984).
24. Mukherjee, *Economic History of India,* 172.
25. Ibid., xix.
26. Ibid., xxiii.
27. Ernst von Weizsacker, Amory Lovins, and Hunter Lovins, *Factor Four: Doubling Wealth, Halving Resource Use* (London: Earthscan, 1997), 50; Vandana Shiva et.al., *Principles of Organic Farming* (Delhi: Navdanya, 2004), 156–163.
28. Lloyd Timberlake, *Africa in Crisis: The Causes, the Cures of Environmental Bankruptcy* (London: Earthscan, 1985), 154.
29. Richard Barnet, *The Lean Years: Politics in the Age of Scarcity* (London: Abacus, 1980), 171.
30. Vandana Shiva et al., *Corporate Hijack of Biodiversity: How WTO Trips Rules Promote Corporate Hijack of People's Biodiversity and Knowledge* (Delhi: Navdanya 2002); Proceedings of EU–India Dialogue on Biosafety and Biotechnology, April 1–2, 2005 New Delhi.

31. Tribal Research and Training Institute, *Malnutrition-Related Deaths of Tribal Children* (Pune, India: Tribal Research and Training Institute, 2002), 4.
32. A. V. Krebs, "The Corporate Reapers: Towards Total Globalization of our Food Supply," in *Sustainable Agriculture and Food Security: The Impact of Globalisation,* eds. Vandana Shiva and Gitanjali Bedi (Thousand Oaks, CA:. Sage, 2002), 187.
33. Charan Singh, *Economic Nightmare in India* (New Delhi: National Publishing House, 1984), 119.
34. Quoted in Oxfam, *Boxing Match in Agricultural Trade: Will WTO Negotiations Knock Out the World's Poorest Farmers?* (Oxfam Briefing Paper no. 32, November 19, 2002).
35. Eric Schlosser, *Fast Food Nation: What the All-American Meal is Doing to the World* (London: Penguin, 2002), 240.
36. Ibid., 241.
37. http://www.organicconsumers.org/organic/corporate120604.cfm
38. http:/Inews:independent.co.uklworld/science_medical/story.jsp?story=453124)
39. Matin Qaim and David Zilberman, "Yield Effects of Genetically Modified Crops in Developing Countries," *Science* 299, no. 5608 (February 7, 2003), 900–902.
40. Jeffrey M. Smith, *Seeds of Deception. Exposing Industry and Government Lies About the Genetically Engineered Food You are Eating* (Fairfield, IA: Yes! Books, 2003), 90–93.
41. Arthur Hocart quoted in Rifkin, *Biosphere Politics,* 39.
42. John Cavanagh and Jerry Mander, eds., *Alternatives to Economic Globalisation* (San Francisco: Berrett-Koehler, 2004), 45.
43. N. S. Jodha, *Market Forces and the Erosion of Common Property Resources* (Hyderabad, India: ICRISAT, 1986) mimeo.
44. Sean P. Murphy and Raphael Lewis, "Big Dig Found Riddled with Leaks: Engineers Say Fixes Could Take a Decade," *Boston Globe* (Nov. 10, 2004).
45. See Chapter Four, Earth Democracy in Action.
46. Steve Chapman, *Takings Exception: Maverick Legal Scholar Richard Epstein on Property, Discrimination, and the Limits of State Action,* Reasononline (http://reason.com/9504/epstein.apr.shtml).
47. Steve Chapman, *Takings Exception,* Reasononline.
48. Charlotte Waters, *An Economic History of England 1066–1874* (London: Oxford University Press, 1928).

49. Jeremy Rifkin, *End of Work: The Decline of the Global Labor Force and the Dawn of the Post-Market Era* (New York: Putnam Books, 1995), 3.
50. Sarah Anderson and John Cavanagh in *Alternatives to Economic Globalisation,* eds. John Cavanagh and Jerry Mander (San Francisco: Berrett-Koehler, 2004), 45.
51. M. K. Gandhi, *Hind Swaraj* (Ahmedabad: Navjivan Press, 1938), 61.
52. Claude Alvares, "Deadly Development," *Development Forum* 11, no. 7 (1973): 3.
53. Peter Kropotkin, *Mutual Aid: A Factor of Evolution* (www.calresco .org/texts/mutaid1.htm).
54. Daniel Fife, "Killing the Goose," *Environment* 13 no. 3 (April 1971): 20–22.
55. Garrett Hardin, "Lifeboat Ethics: The Case Against Helping the Poor," *Bioscience* 24 (1974): 561.
56. Quoted in Vandana Shiva and Mira Shiva, Women, *Population and Environment: A Report for the International Conference on Population and Development, Cairo* (Delhi: RFSTE, 1994), 13.
57. "New Shift in Depopulating Strategy" (1991) in UBINIG, *Depopulating Bangladesh: Essays on the Politics of Fertility* (http:// www.hsph.harvard.edu/Organizations/healthnet/SAsia/depop /Chap8.html).
58. Mathis Wackernagal and William Rees, *Our Ecological Footprint: Reducing Human Impact on the Earth* (Gabriola Island, BC, Canada: New Society Publishers, 1996), 9.
59. Amory Lovins, *World Energy Strategies: Facts, Issues, and Options* (London: Friends of the Earth, 1975), 133.
60. Sale, *Rebels Against the Future,* 39.
61. Mukherjee, *Economic History of India,* 95.
62. Mahmood Mamdani, *The Myth of Population Control* (New York: Monthly Review Press, 1972); Jean Dreze and Mamta Murthi, *Fertility, Education and Development* (London: Suntory Center, LSE, January 2000).
63. M.K. Ghandhi, "Economic Constitution" *Young India* (November 15, 1928).
64. Vijay Joshi, "Myth of India's Outsourcing Boon," *Financial Times,* November 15, 2004.
65. Shri Manila Graham Dog Dijet Papa, "Basic Philosophy and Practices of Our Organization."
66. Ibid.

67. R. Kothari quoted in Vandana Shiva et. al., *Ecology and the Politics of Survival* (New Delhi: United Nations University and Sage Publishers, 1991), 341.

CHAPTER 2: LIVING DEMOCRACY

1. Lee Kyung Hae, *Korea Agrofood* (April 2003).
2. Jacques Berthelot, "The Basic Concepts Used for Agricultural Policies are Tricky: Import Protection is the Least Protectionist Way of Supporting Farmers (1st Part)" *Solidarite,* 2nd World Social Forum, Porto Alegre, January 31 to February 5, 2002 (http://www . abcburkina.net/english/eng_pol-agri/eng_protection.htm).
3. Amy Chau, *World on Fire: How Exporting Free Market Democracy Breeds Ethnic Hatred and Global Instability* (New York: Anchor Books, 2004), 9.
4. Research Foundation for Science, Technology and Ecology (RFSTE), *Globalization and Democracy Report of RFSTE* (Delhi: RFSTE, 1995).
5. Quoted in Vandana Shiva, "Democracy in the Age of Globalisation," (keynote address, National Conference on Grassroot Democracy and Threat to Survival: Agenda for Voluntary Associations and Panchayati Raj Institutions, New Delhi, December 21–22, 1995), 3.
6. http://www.cnn.com/WORLD/9509/india_kfc/
7. See Chapter Four, Earth Democracy in Action, for more details on these movements.
8. Vandana Shiva, *Biopiracy.*
9. Leo Tolstoy, *The Law of Violence and the Law of Love,* trans Vladimir Tehenkoff (Santa Barbara: Conford Grove Press, 1983), 84.
10. M.K. Gandhi, "Equal Distribution," *Harijan* (August 25, 1940).
11. David J Rapport, Bill L. Lasley, Dennis E. Rolston, Managing for Healthy Ecosystems (Boca Raton, FL: CRC Press, 2003), 322; http://healthwrights.org/books/HHWL/HHWLchapt15.pdf
12. Vandana Shiva, *Violence of the Green Revolution* (London: Zed, 1988), 109.
13. Vandana Shiva et al., *Principles of Organic Farming.*
14. Howard and Wad quoted in M.K. Gandhi, *Food Shortage and Agriculture* (Ahmedabad, India: Navjivan Publishing House, 1949), 185.
15. G. Clarke, Presidential Address to the Agriculture Section of the Indian Science Congress quoted in M.K. Gandhi, *Food Shortage and Agriculture* (Ahmedabad, India: Navjivan Publishing House, 1949), 83.

16. Edward Goldsmith, *How to Feed People Under a Regime of Climate Change* (London: Ecologist, 2004), 7.
17. Vandana Shiva, *Yoked to Death: Globalisation and Corporate Control of Agriculture* (Dehli: RFSTE, 2001), 6.
18. Debal Deb, *Industrial vs. Ecological Agriculture* (Delhi: Navdanya, 2004), 57.

CHAPTER 3: Living Cultures

1. His Holiness the Dalai Lama, *The Global Community and the Need for Universal Responsibility* (Boston: Wisdom Publications, 1992).
2. Rudolf Bahro, *From Red to Green* (London: Verso, 1984), 211.
3. http://www.newamericancentury.org/
4. Quoted in Vandana Shiva (keynote address, Women and Religion Conference, Centre for Health and Social Policy, Chiang Mai, Thailand, February 29-March 3, 2004).
5. Quoted in Vandana Shiva, *Tomorrow's Biodiversity,* 131.
6. Quoted in Vandana Shiva, *Globalization, Gandhi, and Swadeshi: What is Economic Freedom? Whose Economic Freedom?* (Delhi: RFSTE, 1998).
7. Maureen Dowd, "Slapping the Other Cheek," *New York Times* (November 14, 2004).
8. http://observer.guardian.co.uklinrernational/story/0,6903, 1075950,00.html
9. "From Hindurva, Advani Takes One Little Leap, Says God Chose BJP," *Indian Express* (November 27, 2004).
10. Dharampal, *Despoliation.and Defaming of India,* 103.
11. Ibid.
12. "Late Rains to Boost Bumper Foodgrain Production," *Economic Times* (November 26, 2004): 3.
13. Vandana Shiva, et.al, *Corporate Hijack of Biodiversity* (Delhi: RFSTE, 2003), 15.
14. WHO, *World Report on Violence and Health* (Geneva, 2002), 215.
15. UN Convention on the Prevention and Punishment of the Crime of Genocide (http://www.unhchr.ch/htmlf menu3/b/p_genoci.htm).
16. Veeresh Committee, *Farmer Suicides in Karnataka: A Scientific Analysis* (Bangalore: Government of Karnataka, 2004), 113.
17. Ibid.
18. Vandana Shiva, *Towards People's Food Security* (Delhi: RFSTE, 1995), 16.

19. Vandana Shiva and Kunwar Jalees, *The Mirage of Market Access: How Globalization is Destroying Farmers Lives and Livelihoods* (Delhi: RFSTE, 2003), 47.
20. Vandana Shiva, *Staying Alive: Women, Ecology and Development* (London: Zed Books, 1989), 118.
21. Associated Press, "Over 60 mn Women Have Fallen Victims to Sex Discrimination" *The Indian Express* July 24, 1997).
22. Sarala Gopalan and Mira Shiva, eds., *National Profile on Women, Health and Development* (New Dehli: VHAI, WHO, April 2000), 226.
23. R. P. Ravindra, *The Scarcer Half* (Bombay: CEO, 1986).
24. Lakshmi Lingam, "Sex Determination Tests and Female Foeticide: Descrimination Before Birth," in *Understanding Women's Health: A Reader,* ed. Lakshmi Lingam (Delhi: Kali for Women, 1998), 209–218.
25. Mira Shiva and Ashish Bose, *Missing: Mapping the Adverse Child Sex Ratio in India* (United Nations Population Fund, October 2003), 3.
26. Vibhuti Patel, "Sex Selective Abortions: Pre-Birth Elimination of Girls" in *Health for the Millions* issue on "Population and Development" (Delhi: Voluntary Health Association of India, August–November 2004).
27. Pamela Philipose, "Where Is the Girl Child?" *Indian Express* (April 15, 2001).
28. Ministry of Health and Family Welfare, "Missing Girls: A Case Study From Delhi" in *Missing Census of India* (New Delhi: Ministry of Health and Family Welfare and UNFPA, 2003), 2–11.
29. Ministry of Health and Family Welfare, "Missing Girls," in *Missing Census of India* (New Delhi: Ministry of Health and Family Welfare and UNFPA, 2003), 1.
30. T. K. Rajalakshmi, "Crime Capital" *Frontline* 19, no. 25 (December 7–20, 2002).
31. Mira Shiva and Ashish Bose, *Darkness at Noon* (New Dehli: VHAI, 2003).
32. Samuel P. Huntington *The Clash of Civilizations and the Remaking of World Order* (London: Simon & Schuster, 1997), 20.
33. Walt Martin and Magde Ott, *Albert Schweitzer: Reverence for Life* (Unpublished manuscript), 3.
34. http://www.civworld.org/declaration.cfm

CHAPTER 4: Earth Democracy In Action

1. Research Foundation for Science, Technology and Ecology, *Basmati Biopiracy* (Delhi: RFSTE, 1999), 1.
2. CSS Haryana Agricultural University, Hissar.
3. http://www.iraqcoalition.org/regulations/20040426_CPAORO_ 81 _Patents_Law.pdf
4. International Coalition on the Future of Food and Agriculture, *Manifesto on the Future of Food and Principles of Earth Democracy* (Cambridge, MA: South End Press, 2005), 7–8.
5. Ibid., 24–27.
6. Women and Water Conference (Doon Valley, February 2004).
7. In the High Court of Kerala, Judgement by Mr. Justice Balakrishnan Nair, Tuesday 16th December 2003, WP(C) No. 34292 of 2003 (G), 23.
8. Sureshwar D. Sinha, *Quenching Delhi's Thirst Locally: Rejuvenating the Yamuna and Reviving Local Water Sources* (Delhi: RFSTE, August 2003) 37.
9. Shekar Singh and Pranab Banerjee, *Large Dams in India: Environment, Social and Economic Impacts* (New Delhi: Indian Institute of Public Administration, 2002), 18.
10. Vandana Shiva and K. Jalees, *Ganga: Common Heritage or Private Commodity* (Delhi: RFSTE, 2004).
11. *World Bank Paper No.* 386 (World Bank, 1997).
12. Vandana Shiva and K. Jalees, *River Linking Project* (Delhi: RFSTE, 2004), 1.
13. Ibid., 2.
14. Vandana Shiva and Gunjan Mishra, *Ken-Betwa Link: A Social and Ecological Assessment* (Delhi: RFSTE, unpublished report 2004), 1–6.
15. David Pearce, "The Future of the Earth" (European Acadamy of Otzenhausen, Germany, March 2005).

INDEX

Y

Z

Photo by Kartikey Shiva

About the Author

Vandana Shiva is a physicist, world-renowned environmental thinker and activist, and a tireless crusader for economic, food, and gender justice. She is the author and editor of many influential books, including *Making Peace with the Earth, Soil Not Oil, Staying Alive, Stolen Harvest, Water Wars,* and *Globalization's New Wars*. Dr. Shiva is the recipient of more than twenty international awards, among them the Right Livelihood Award (1993); the John Lennon-Yoko Ono Grant for Peace (2008); The Sydney Peace Prize (2010); and the Calgary Peace Prize (Canada, 2011). In addition, she is a board member of the World Future Council and one of the leaders and board members of the International Forum on Globalization (whose other members include Jerry Mander, Edward Goldsmith, Ralph Nader, and Jeremy Rifkin). She travels frequently to speak at conferences around the world.

Photo by Kartikey Shiva

TITLES BY VANDANA SHIVA

available from North Atlantic Books

Soil Not Oil

978-1-62317-043-1 (print)
978-1-62317-044-8 (ebook)

Earth Democracy

978-1-62317-041-7 (print)
978-1-62317-042-4 (ebook)

Seed Sovereignty, Food Security

978-1-62317-028-8 (print)
978-1-62317-029-5 (ebook)

Staying Alive

978-1-62317-051-6 (print)
978-1-62317-052-3 (ebook)

Biopiracy

978-1-62317-070-7 (print)
978-1-62317-071-4 (ebook)

Who Really Feeds the World?

978-1-62317-062-2 (print)
978-1-62317-063-9 (ebook)

Water Wars

978-1-62317-072-1 (print)
978-1-62317-073-8 (ebook)

North Atlantic Books

www.northatlanticbooks.com

North Atlantic Books is an independent, nonprofit publisher committed to a bold exploration of the relationships between mind, body, spirit, and nature.